ID#969 *Spirit* and ID#1042 *Legion* powering through calm, blue water in the Moray Firth.

Contents

Foreword

If one was to invent a creature that would appeal to the human aesthetic, then there are some physical features one might naturally look to include. It's very likely it would live on the land and have physical features with which we could easily identify. These would be likely to include forward facing eyes that lend themselves to binocular vision and prominent ears on the sides of its head (such as are found in the ever-popular meerkat). It would probably have hands that look like our own such as those of a chimpanzee or gorilla and probably have the capacity to sit upright as we do. We are most *unlikely* to conjure up an animal that lives in the sea, breathes through a hole in the top of its head and whose arms and legs have evolved into fins and flippers. It is equally unlikely we would create a being that was able to 'see' with sound and has a complex language made up of clicks and whistles, many of which are beyond the range of human hearing. In fact, it would be hard to invent a creature more *unlike* a human being – or our image of attractive or cute – than is a dolphin.

Yet these creatures occupy a uniquely treasured place in our psyche and affections and when you read this remarkable account you will, I am certain, understand why that may be. Charlie's relationship (and a relationship is precisely what this book recounts) with the dolphins of the Scottish east coast is as respectful as it is heartfelt. Whilst his unique observations and accounts are based on robust techniques, and do not slip into mawkishness, you immediately sense his affinity with these creatures and his lifelong passion to understand and interpret their behaviour and family life. And perhaps herein lies the key to the popular appeal of the species.

Whilst we are undoubtedly visual beasts, drawn and affected by what we see, dolphins touch something deeper within our being than appearances alone might suggest. Of course, we cannot truly know what it is that they think or feel, but watching their behaviour, their reaction to each other and to other creatures (including ourselves) we sense the presence of a sentient creature – one which is completely at home in its element, which effectively manages the politics of its society and which, we are increasingly certain, has a language that communicates much more than the basic necessities of existence. The physical and behavioural barriers between our world and theirs are as great as is our desire to bridge them, and yet for thousands of years, many cultures have held dolphins, and especially bottlenose dolphins, in the highest esteem. Recent scientific research has leant further credence to the idea that dolphins and other cetaceans are self-aware and as such are 'non-human persons'. This has far-reaching consequences since, if it were to be fully embraced by international laws, the killing, incarceration (being kept in captivity to entertain us) or other infringement of their personal rights would be a crime. Charlie would argue that such treatment of these extraordinary creatures is already a crime, regardless of human law and, once you have read this book and marvelled at the extraordinary images within its pages, I have no doubt you will agree.

Simon King

Introduction

The title of this book 'On a Rising Tide' is a reference to the optimal time tidally, in the author's opinion, for wildlife enthusiasts to start watching from land for a sighting of the resident 195 or so Bottlenose dolphins that can be seen around the shores of the Black Isle, near Inverness in Scotland. They can be seen at any time of year but more often and much more reliably during the spring and summer months. Shortly after low tide, as the waters begin to rise (or flood) again after six hours or so of falling (or ebb) tide, day or night, between late spring and early autumn, resident dolphins start to patrol certain areas of the Inner Moray Firth to hunt for seasonal migratory salmon that are coming back from the open sea to the nearby rivers to spawn. These magnificent fish have to navigate past some difficult and often dangerous obstacles to get back to their own rivers, not least of which can be oceanographic bottleneck areas like the entrance to the Cromarty Firth, the Chanonry Narrows (where Chanonry Point is my 'office' for a lot of the year) and, nearest to the city of Inverness, the Kessock Channel, part of the Beauly Firth where the sea is deep and narrow between land masses. Places like these are where large predators can gather to ambush their prey in the fast moving tidal currents – a clever hunting method that we call 'against current foraging'. Many of the dolphins in this rare, very northerly population have learned this technique traditionally down many generations from their mothers and have perfected it, as you will see later in this book, to deadly and spectacular effect. Dolphins from this population can be regularly sighted all the way from Golspie on the Sutherland coast southwards to the Tay and Forth estuaries and nowadays over the border into north-east England. However, most of my observations about sightings and times and stories about the dolphins will be from the very Inner Moray Firth at Chanonry Point and the Kessock Channel and also from being out on the water on board research vessels or tour boats around the Cromarty Firth and beyond.

I hope that you enjoy the photographs in this book. I'm getting a bit 'arty' in my old age and have chosen a set of images that are, in my opinion, a wee bit different from being simply 'leaping dolphins' of which there are many, many thousands of frames taken by dolphin watchers and photography enthusiasts every year. I also hope that you like the personal stories and narrative about my twenty or so years (so far) of encounters with these wonderful, big, highly intelligent and sentient souls. Forgive me if I ramble a bit occasionally during parts of this book.

When you are trying to describe what dolphins are *actually* doing, and not what we *think* they might be doing, while trying to put across the sheer excitement and complex emotions that flood through you when watching or studying these dolphins – being succinct sometimes just isn't that easy. Every dolphin in this population that has been recorded photographically by Aberdeen University has an identity number of its own. For instance ID#1025, who has been named Charlie after me – a great honour! But for the purposes of this book I will try and use the names that many, but not all, of the dolphins have been given, rather than numbers that can sound a bit cold.

Chapter 1

Early memories of dolphins

In the late nineteen sixties and early seventies our family would come on summer holidays to the beachside caravan site in the lovely fishing village of Portmahomack in Easter Ross. My father had a small sailing dinghy (a GP 14 as I recall) that we used a lot during the often sunny but breezy weather, sailing up and down this part of the Dornoch Firth (an offshoot of the main Moray Firth). My earliest recollection of encountering these Bottlenose dolphins is that groups of them would sometimes come and bow-ride with us as we sailed along at a good pace. I was already interested in sea-life by then and knew that they were probably Bottlenose dolphins, but they didn't look quite the same as the slender, light grey ones that I saw in my nature books – these dolphins were much darker and bigger. I asked some of the local fishermen about them and they told me that the dolphins lived around here and of the great respect that they had for them. Their grandfathers and great-grandfathers had liked and admired them too and would never do them any harm, being very careful when working out at sea if dolphins were close to the fishing boat. I mentioned that some of them were nearly as big as our fourteen-foot-long sailing dinghy and was told by one very knowledgeable old chap that because these dolphins lived in very chilly waters in the north-east of Scotland, they probably needed more fat to keep out the cold. That was why they were so big compared to the ones I had been looking at in sea-life books. He also stated that if they, the fishermen, saw a group of dolphins and accompanying gulls while out at sea, then there was a good chance of catching some of the fish that the dolphins were after – a good working partnership! I have found over the years that most of the local fishermen have a practical and non-fanciful understanding of wildlife and nature. They have an inherent wisdom and sometimes a brutal honesty that can bring you down to earth every so often, so it is always worth spending time chatting to them.

Opposite: five sub-adult male dolphins on their boisterous travels in the Moray Firth.

Chapter 2

Winding forward a few years

So – moving forward to the late nineteen eighties and I find myself up in the Moray Firth living and working in the coastal town of Cullen. Almost as soon as I am settled here I realise that I can regularly see the same species of dolphin from my living room window as I did in Portmahomack all those years ago; I get the feeling that something is bringing these amazing mammals and me back together after twenty or so years. Luck or coincidence? Serendipity maybe? – you may choose for yourself. I was absolutely thrilled by the way that some of the dolphins would come very, very close to the rocky headlands where I would often sit as they travelled past the wide mouth of Cullen Bay. Time would be totally forgotten as I sat on the rocks just past the harbour, or round the coastal walk nearer Sunnyside Beach, entranced by the presence of these very active, huge animals. I also became good friends with a local fishing skipper, Chick Anderson, who had a small but very serviceable clinker-built old boat that he let me use at any time (in return for repairing the stubborn, recalcitrant, single-cylinder Lister inboard engine). I spent a lot of my spare time out in Cullen Bay and beyond, waiting on a group of dolphins that seemed to come past at roughly the same time most days. Sometimes the dolphins would simply be transiting – that is, going from A to B in a determined fashion but at other times they would be actively hunting for shoaling fish of various species. That was exciting, as the dolphins would come right under the boat and sometimes surround me – fish going all over the place!

The remarkably scarred dorsal fin of ID#573 *Denoozydenzy* breaks the surface only a metre or so from the author's camera.

It was during one of these many 'close encounters' that I realised that the big, pointy dorsal fins on the dolphins' backs looked a bit different from each other with chunks missing from some fins, while many individuals had long, white-coloured scratches about the fin, tail flukes and body too. After a few encounters I was positive that I could tell a few of the regular dolphins apart as they were so different in appearance. In some of the slightly blurry photos that I was starting to take on a small compact film camera I could see the same scratched fins appearing again and again. Sadly I don't have many of these early photos now as during a house move a few boxes went missing. On one occasion, during a very hectic

bout of activity by the dolphins around the boat, the camera strap came adrift from the camera body. All I could do was watch as the camera, complete with 34 or so fantastic dolphin-hunting-mackerel photographs headed downwards to the sea bed. Little did I know that around the same time, marine mammal scientists from Aberdeen and St Andrews Universities, who were working initially on seal research, were starting to identify individual dolphins around the Cromarty and Inner Moray Firth area and were indeed seeing the same dolphins again and again. Out of the blue, a geographically isolated and pretty rare resident group of dolphins was slowly but surely being discovered.

Above: ID#815 *Prism* looking right at the author as he surfaces to breathe.

Opposite: ID#1025 *Charlie* (nearest) with pals ID#1018 *Bodhi* to his left and, re-entering the water behind him, ID#989 *Mini Fin*.

Chapter 3

Old friends

A chance encounter at an event in the Town Hall at Cullen got me in touch with a fledgling group of individuals who were interested in studying the dolphins and I came to realise that it wasn't just me who was bewitched by them. There were like-minded individuals all along the Moray coast, watching out from the coastal villages and towns and soon I was involved on the committee of Friends of The Moray Firth Dolphins with the irrepressible Pete MacDonald at the helm. Over time the group blossomed and a very efficient network of watchers developed, giving early warning of sightings of dolphins, roughly which direction they were travelling in and an idea of numbers and social make up. Many a happy hour was spent watching with other group members from the back of Pete's family house at Findochty (which faced the sea from an elevated position). Although the sightings of dolphins were brilliant, the company and the laughter were extra special, especially as the sun was going down and the colours across the Moray Firth turned to gold. After a while, the group had enough funding to buy a small boat that Pete looked after and ran from Findochty (pronounced 'Finechty') for the group members plus friends. Many happy days were spent out at sea in the company of these wonderful dolphins… the human company wasn't that bad either and many an adventure was had and many a (tallish) tale told at parties over the years of dolphin encounters. I raise a glass to Pete and Lyn, Tony and Janey, Duncan and Jean, Rodger, Tess and a host of others for raising awareness of just how special these dolphins actually are. This group played a significant part in having the national importance and value to the area of the dolphins brought to the fore.

Chapter 4

The green shed

Although no longer wholly green in colour, this is the affectionate name given by locals, visitors and dolphin-watching enthusiasts to the Dolphin & Seal Centre near the Kessock Bridge. Situated about one hundred feet above sea level and overlooking the Kessock Channel (part of the Beauly Firth), this little wooden structure started off in 1995 as an experiment by the local Ross and Cromarty District Council. The idea was to see if there would be sufficient demand for a much bigger, full-time and professionally manned interpretive centre, with marine mammals like the resident dolphins, seals and sometimes otters being visible from the panoramic viewing window. With expert scientific input from Aberdeen University and support from Scottish Natural Heritage, this tiny interpretive centre includes the unique attraction of a live hydrophone system so that the public can hear underwater sounds as they happen. The Dolphin & Seal Visitor Centre opened to the public and is still open seasonally to this day, run and managed nowadays by my colleagues from Whale and Dolphin Conservation.

I joined the seasonal interpretive staff in the early spring of 1996 after my friend Tony Archer had relinquished his position to go and set up the Moray Firth Wildlife Centre with his wife and family at Spey Bay (now the WDC Scottish Dolphin Centre which I helped manage for a year). With enthusiastic colleagues I spent the next five seasons in this lovely, creaky little wooden building, dealing with the many thousands of visitors that passed through the doors every year, nearly all looking for information about how and where to see our dolphins – who were becoming very well-known natural superstars through various television programmes. I also learned a lot from the Aberdeen University and other affiliated marine biologists and zoologists from around the world, who were utilising the hydrophone system and sometimes the wall-mounted video camera for studying not only the local Bottlenose dolphins, but also the resident Common (harbour) seals that could be seen and heard swimming around and hunting for fish on a regular basis in the waters below us.

It was during my work at North Kessock that two very important things happened to me… and I need to make sure that I get them in the right order of importance here, folks! Firstly I met Susan, my lovely future wife and secondly I struck up a working relationship with staff from the UK-based marine charity Whale and Dolphin Conservation Society who would pop into the Centre now and again. Little did I know that after doing some 'Dolphin diary' articles for a year or two and giving WDCS a few dolphin pictures for their magazines and brochures, that I would be offered a full-time post as their field officer. My main responsibility is to service the 'Adopt a Dolphin' programme with up-to-date information and photographs of six particular dolphins that the public can 'adopt' and so support the charity financially. It is a job that I still love all these years later and although sometimes during the summer the working day never seems to end, I wouldn't exchange it for any other job on the planet, or for that matter any other post within what is now Whale and Dolphin Conservation (WDC). Mind you, I would really love to spend a few months across in Canada photographing the beautiful Orca (Killer whales) and helping with the research that WDC and Orcalab is involved in with these biggest and most awe-inspiring members of the dolphin family.

Chapter 5

Photographing dolphins

As a technical exercise, photographing any large, fast-moving wild animal that lives in the sea is going to have its challenges – not getting yourself and your camera equipment soaked in sea water is the most obvious thing for a start. Given that you can stand on a nice beach with dolphins leaping around only a few metres from the shore up here in the Inner Moray Firth, is it really necessary to clamber on board a boat, get cold, wet, possibly bruised and go

out on the chance of maybe seeing something close enough to photograph? The answer is simple really – try both methods and see how you get on. I love doing both as it gives me a chance to see and experience these dolphins going about their daily lives from completely different aspects and angles. There are plenty of land-based stories in this book but I have tried to include a few adventures out on the big blue sea as well just for balance.

Above: ID#990 *Flake* nearest the camera and pal ID#818 *Yin-Yang* in the Inner Moray Firth.
Page 14: ID#990 *Flake* and pals in the Inner Moray Firth with an Oil Tanker in the distance.

Over the years I have built up some great relationships with local commercial tour boat owners and operators who are willing to put up with my non-stop chatter and running commentary whenever a dorsal fin or two appears – in exchange for me giving their guests a bit of information and chat about which species or the individual dolphins that we might be seeing. I'm always fairly generous to these hard-working and dedicated business people with my photographs taken on-board their vessels for brochures and websites if they don't have any good ones of their own. It's always nice to be nice and from a purely selfish angle it's another chance to be mobile out at sea as you never know what else might turn up to photograph as well as dolphins.

I am also in the extremely fortunate position of being regularly invited by the research staff of Aberdeen University's Lighthouse Field Station at Cromarty to join them on whole-day dolphin identification surveys on board their lovely rigid inflatable boat called the *Rona* as second photographer, Barbara Cheney being the dolphin Photo ID Officer at the Lighthouse and principle photographer. These trips are very, very special to me as this isn't just opportunistic, boat-based snapping away at dolphins with a camera. The terms of the research license mean that we are working under strict guidelines and protocols as these dolphins are highly protected, both under national and European law. This means that we have to be extra careful about how dolphins are approached and the way those encounters are managed.

One of the upsides of these strict 'rules of engagement' is that we are allowed to spend a little bit more time with groups of dolphins than other boats in order to obtain dorsal fin identification photos – images that are fundamental in identifying the individual members of this dolphin population. The spin-off data that comes from analysing these photographs can be used to underpin a huge amount of fascinating research and give the evidence needed for recommendations for important government conservation reports. Eight hours or more bouncing around on a boat trying to take sharp, well-focussed pictures of dolphins can be very hard work both mentally and physically, believe me. The last all-day survey I was lucky to be on had me taking well over 2,500 images, with three big memory cards full to capacity and a lot of work for the next few days going through them all, sorting out the good from the bad. At least sitting in front of the big computer monitor the next day drinking lots of coffee doing a ruthless edit of these photos gave my legs a much-needed rest!

I'm not going to bore you to death with a lot of technical camera jargon about F stops and ISOs throughout this book (that kind of stuff is for another book altogether) but if you are interested, my personal preference is to set my DSLR cameras up (main camera and backup) in the same way pretty much all the time. Manual mode, 1600th second shutter speed, F6.3 aperture and fully auto ISO (not all cameras have this facility) with +2/3rds stop exposure compensation. That setup seems to work for me with only minor changes made in exceptional circumstances – when you have loads of dolphins around then you really don't have much time to sit or stand about fiddling with settings for too long. To put it a bit more simply, the most important thing to remember when trying to photograph dolphins, either from land or on board a boat is a combination of fast camera setting and your own reaction speed. The faster the better in both cases and a bit of practice doesn't go amiss either, even before you go anywhere near the sea.

Lens-wise, if you have a camera where you can swap lenses then I would say that a 70-300mm is light enough and portable too, and would do you nicely for land or boat. Doing justice to dolphins from land may require a longer focal length, something like a 100-400mm zoom lens for example. If your camera has a 'Sports' setting then try it out on something like your pet dog running around, or your own kids if you are so blessed – either will do nicely to get used to capturing moving objects in sharp focus. If there is a setting on your camera that allows exposure compensation I advise overexposing a little: the sea can be very bright and the camera might try to compensate by darkening it, making your dolphins look a bit dark too. Once you are confident about using the camera then please, please remember to have spare batteries and memory cards with you. Ensure the batteries are fully charged and the memory cards freshly formatted (cleared of all pictures) so you will be ready to take possibly many hundreds of dolphin photos. There is nothing so frustrating as having a group of dolphins right beside you and your battery goes flat or the memory card fills up. And don't forget a nice, clean, soft lens cloth for the lens front in case it gets splashed. As the old Scouts motto goes… be prepared.

Page 15: ID#105 *Sundance* and pal ID#907 *Beatrice* breaching from the water at Chanonry Point.
Opposite: ID#11 *Muddy* (tallest fin in the group) with associates in the Moray Firth.

Chapter 6

An early start

It is just after four-thirty in the morning as I haul my camera gear out of my old Honda four-wheel drive. I'm parked on the old, disused sandstone slipway at Chanonry Point – my usual 'spot' as I'm going to be there for the next six hours or so. This frees up a valuable designated parking space later in the morning for a member of the public at an ever more crowded Chanonry car park. Low tide isn't until just after 5.00am but these are good, powerful spring tides and it won't take long before the tidal flow starts to reverse itself, so I'm being cautious by getting here in good time. As I trudge slowly down the shingle beach, it's just beginning to get light and the huge bulb element is still glowing on and off inside its lens housing at the Chanonry Ness Lighthouse. I have my big carbon fibre tripod over my left shoulder, connected tightly to which is the huge 600millimetre Canon lens that is in turn coupled to the big, chunky professional 1D MKIV digital camera body. I'm intending to do nothing but Photo ID with the dolphins' dorsal fins today, hence my biggest lens being with me. But just in case things get a bit lively, I also have my backup camera body with a great quality and versatile 70-200mm F2.8 IS zoom lens attached to it plus a pro-grade video camera tucked safely in my metal-framed rucksack. This has a dinky and very, very handy inbuilt seat – great for taking the weight off your legs for a while if you are down here for a five- or six-hour shift. There is hardly a breath of wind and nothing is stirring in the water, hardly any birds around and best of all – nobody else! For a wee while I have the luxury of having Chanonry Point to myself. Don't get me wrong, I'm not the least bit antisocial but having some quiet time alone and being able to concentrate and set up my tripod and camera gear properly and do my pre-shoot checks without distractions or interruptions is a rare thing. For a while at least it's absolute bliss. There is also something of a selfish pleasure of having 'my' beach to myself for a short time.

Although it is getting a bit lighter, it's not exactly what I'd call a bright morning, with a steely grey sheen covering the water but without any sunny glare coming from the rising sun. However, it is a particularly good morning for looking away into the distance to see if there are any dorsal fins becoming visible. The tide has, as I suspected, turned around very quickly and I'm detecting a distinct 'left to right' flow in the water rising in front of me. Something catches my eye across the water at Fort George, about a kilometre away – a disturbance in the water, a slight splash but I can't see anything else that would give me a clue as to what it was – it might have been a dolphin or perhaps a lively seal. I don't have to wait long before I see, coming right towards me from the same direction, a huge dorsal fin breaking the surface and a loud 'whuff' of an exhalation of breath reaches my ears as my first customer of the day arrives. It is none other than the mighty Nevis who, apart from being a great favourite of mine, is one of the oldest, biggest, and most senior dominant males in the population. He has a snowy white de-pigmented top to his wonderfully battle-scarred fin and was one of the six WDC Adoption dolphins that I study and photograph a little bit more than the other individuals. He swings his huge bulk round with surprising agility into the ever more powerful tidal current, vents loudly three times and then his towering, craggy dorsal fin vanishes beneath the surface.

While immersed in the silty, murky water, Nevis is sending out loud streams of echolocation clicks generated from the air sacs within his blowhole and is focussing these trains of clicks through the acoustic lens or 'melon' near the front of his head. He receives the clicks back through his bottom jaw, which is acting like a super-sensitive radio aerial as he free-falls down through the cold, turbid water to about fifty feet or so beneath the surface, trying to detect his breakfast swimming towards him. It is almost seven minutes before I see him again – an amazingly long dive for a dolphin to do here. Normally, a minute or so is all that is needed to scan and detect prey rushing towards the dolphin. I thought that he had ambled off underwater and I hadn't seen him go; he was down for so long and I have never timed a dolphin being immersed for as long before or since. It is still very calm and quiet and there are no other dolphins around so I have the time to set up the video camera and manage to get a little low-light footage of the big guy mooching about.

With hindsight I am extremely glad that I did, as, sadly, this was to be the last time that I saw Nevis alive. A month or two later after not having seen him around (which was most unusual), I received a phone call while on my way back from a trip to the west coast of Scotland filming Common dolphins around the Small Isles. It was from my friends at the Lighthouse Field Station at Cromarty to say that the body of a big, very old male dolphin had been found on the beach at Balintore, just north of the Cromarty Firth – a frequent 'wash-up' location for dead seals, porpoises and dolphins due to the way that the tidal currents work in this part of the coastline, and that it was, no doubt about it, Nevis. It was the end of an era, so to speak and I was very subdued and rather depressed for a few days knowing that his familiar, massive presence – whether hunting and socialising at Chanonry Point or the Kessock Channel, would be there to enjoy no more. The death of any of our dolphins is bad enough but there are some extra complications for me and my head office support staff where the WDC 'Adopt a Dolphin' individuals are concerned. It means having to undertake the sad task of informing all the adopters, dealing with the fallout from that and then offering an alternative dolphin – never easy and always very emotional. People get extremely attached to their dolphins – and that includes me too.

Opposite: ID#36 *Nevis* in full hunting mode, launching a salmon into the air in rough seas at Chanonry Point.

Chapter 7

There be giants

Although not actually about our dolphins as such, I'd like to mention a late-in-the-day adventure that I had a few years ago, so please forgive me as I go a bit 'sideways' here. A series of frantic phone calls (some from British Divers' Marine Life Rescue, some from our then science department head, Mark Simmonds) came through to me one day very close to Christmas Eve about a large whale, thought to be a big old male Sperm whale that was seemingly trapped in a sea-loch near a fish farm over on the Isle of Skye. Since I was 'local', could I pop over and have a look and ascertain what sort of physical condition it was in and advise as to what to do next? Two-and-a-half hours driving later I am on my second home – the Isle of Skye – meeting up with Armadale tour boat operator Peter Fowler who has very kindly agreed to take me out on a small inflatable in rapidly worsening weather to try and get a look at this huge animal. Sure enough, it is pretty close to the fish farm but is also moving around a fair distance away from it too, so isn't entangled. The wind is by now whistling down the mountains and Peter and I get launched and start heading slowly and carefully towards the whale, so as not to startle it. The big animal appears to be confining itself to a particular area of the loch, swimming slowly and deliberately and looking very thin, with a great dip visible behind the massive spermaceti organ on the head. The whale has clearly used up a lot of its fat reserves and is running on empty. I get the camera fired up and although the light has almost gone, I turn up the ISO and start blasting away, trying to get as many useful pictures of its physiology as I can for assessment purposes. The next forty-five minutes or so are like a Wild West rodeo with Peter and I getting thrown up and down but eventually we decide that it's time to leave this gentle giant alone and get back to shore.

Thanking Peter profusely for the use of the boat and waving him off, I stopped for a quick chat with Mike Rae, a professional photographer who happened to be in the area and had some photos of the whale. I had to get going back to Inverness so I said my goodbyes and started the long, dark drive home. I got home by around 11.30 at night and then had to start downloading the pictures and e-mailing them to numerous expectant whale scientists and rescue co-ordinators. So my Christmas Eve rolled over with a computer mouse in my hand instead of a glass of wine and at around 1.30am I finally shut the computer off, headed for bed and spent the rest of the night trying to sleep but still having the feeling of being tossed up and down on the tiny inflatable. A few days later the whale found its way out of the sea loch and headed north to colder, deeper water where its main prey species, big squid, lives. Hopefully, a happy ending.

An underweight adult Sperm whale (*Physter macrocephalus*) at the surface to breathe in Loch Ainort, near Luib on the Isle of Skye.

Chapter 8

Colourful mornings

I'm sitting on the old slipway at North Kessock on a crisp, very cold and still, mid-January morning with the tide starting to rise. The morning sun is rising too, turning the water a beautiful golden-bronze hue. It is not a time of year when dolphins are around in any great numbers but, sure enough, in the distance I can hear the familiar exhaling of a dolphin somewhere within the ever-brightening pool of light over near the end of the Caledonian Canal, about five hundred metres from where I am sitting. A cloud of vapour from the dolphin's next warm breath hangs in the chilly air, the dolphin much closer now and easily focussed on, allowing me time to fire off a sequence of frames. Checking the photos on the screen on the back of the camera I am quietly pleased with the way that the dolphin seems to be sitting in a pool of liquid gold with the lighter chimney of breath vapour hanging above the whole scene. A few hundred metres away and closing in fast is another, slightly smaller dolphin, with a much more pointed dorsal fin... Kesslet, and her young son Charlie, who has been away examining something further up the Firth and who has been whistled back to his mum's side, have arrived for a morning's hunting. At this time of year there could be the very early start of the migratory salmon but as it has been so cold in the Firth there is also the possibility of some winter herring being available – a species that used to be fantastically abundant here until, as usual, commercial greed and overfishing in the 1960s and '70s stripped out nearly all the breeding stock. Very, very slowly however, there have been signs of a recovery in shoal sizes and these tasty and very nutritious fish could once again be on the dolphins' winter 'menu', as seems to be the case this morning.

ID#433 *Kesslet* and ID#748 *Trail Scoop* surfacing to breathe near Chanonry Point.

Young Charlie, bless him, although being every bit as fast, agile and nimble as his very svelte and manoeuvrable mother, tends to let his youthful enthusiasm dominate his hunting tactics. Instead of carefully circling the shoal of herring diagonally opposite Kesslet, closing the ball of fish tighter and tighter until the fish stupefy themselves through lack of oxygen – he goes charging right in amongst the small, silvery fish, scattering them in all directions and perhaps managing to grab just one for his trouble. Through the big telephoto lens I can see and almost feel Kesslet's sense of exasperation and frustration as the shoal of fish will have to be herded into a suitable area all over again. But sure enough, after a while Charlie does as he is told and together, after a few hours they manage to get a good, hearty meal that will hold body and soul together until the next incoming tide. To be able to witness this quite rare spectacle in such lovely light was one thing, but you can easily forget about the bitter cold while watching dolphins for a few hours. I was so stiff and numb that I had to go back up the slipway and get into the car to try and warm up for a while before I started to dismantle the camera gear and pack it away. 'Dropped camera and lens because of frozen hands' would be a new and slightly novel entry for me on the insurance claim form, I think!

Almost exactly a year later from the Kesslet and Charlie herring hunt episode in the Kessock Channel, I was invited out on a friend's lovely yacht, moored at Inverness Marina for an early morning trip out to the Cromarty Firth and back. As we were heading back in towards Inverness on the return journey, in worsening conditions near the Munlochy Shoals marker buoy, Brian and I came across a big group of dolphins that were hunting as a team, and it seemed as though herring were once again very much on the menu. Around twenty or so of our local dolphins, including Moonlight, Sundance and Mischief (three of the WDC 'Adopt a Dolphin' stars) were rocketing around in a closed circle that must have measured a hundred metres or so in diameter. As we came up to where they were, half of the group broke away and came right up beside us – Sundance and Mischief were nearest the boat's hull and then they all dived together and vanished into the chilly, slate-grey water only to emerge at the other side of the boat with herring flying all over the place. For ten or so very uncomfortable minutes and hanging on to the rigging for dear life and getting covered in spray, I fired away with the big Canon 1DX camera at twelve frames a second to get some dorsal identification shots (rare at this time of year). And simultaneously I'm thanking Canon quietly under my breath for having such effective weatherproofing on their camera bodies and also for the nice, deep lens hood on the 70-200 F2.8 IS MKII lens as very little salt spray hit the lens glass at all. The photos that I managed to get that morning are astonishing in my opinion, considering the grey, dark conditions which really pushed the camera, lens and me to the limit – all of this before mid-day too, and very little in the way of breakfast either! A few years ago, even with the top-end gear that I have to use, I simply could not have obtained these top-grade photos in similarly poor conditions but with today's technology, the boundaries of photography are getting pushed back more and more, to almost mind-bending possibilities.

Opposite: top left: ID#748 *Trail Scoop* leads the way at sunrise with ID#1020 *Idris* then ID#578 *Chewbacca* and her calf. Top right: ID#31 *Rainbow* and calf in the bronze water at sunrise. Bottom left: ID#1025 *Charlie*. Bottom right: his mum ID#433 *Kesslet* venting in the chilly morning air.

Opposite: ID#433 *Kesslet* breaching in the winter sunshine.
Above: ID#105 *Sundance* surfacing at dawn, Chanonry Point.

Some of my early mornings can be a little more filled with nerves than others, especially when there are television crews wanting to come with me to film the dolphins. Helping out behind the scenes as a species 'expert' or technical advisor, or being in front of the camera doing a piece about dolphin behaviour and general dolphin facts and figures, can leave you wanting to lie down in a darkened room after a short while. I have worked with many big names in the world of wildlife and conservation over the years and it is always nice to be sought out personally to be involved in such productions. The very way that filming a wildlife programme works doesn't always make sense at first – things are often shot in reverse order and then edited and stitched together later. My biggest worry is that we don't see any dolphins at all or have just missed them or the weather changes, disrupting the continuity of the shots. Thankfully, most film crews are well used to getting on location very early for 'golden hour' footage of sunrise/sunset and are also ready for circumstances changing throughout the day. Having to re-shoot scenes again and again happens as a matter of course.

Above: ID#1025 *Charlie* bow-riding with a yacht in the low, winter sunrise in the Kessock Channel.

Opposite: ID#580 *Moonlight* surfacing to breathe, early morning at Chanonry Point.

Chapter 9

Springwatch

In June 2008 I was asked to help out as a spotter and technical expert with one episode of the BBC *Springwatch* series and, as usual with these big outside broadcast programmes, there was a lot of pre-planning to do and more than a few e-mails, phone calls and Skype video calls before things actually started to roll. I was

delighted that the host of the outside broadcast part of the show was to be Simon King, the great wildlife cameraman, film maker and naturalist who has very kindly written the foreword for this book. The outside broadcast unit is a whole 'village' in itself and took over a lot of the car park at Chanonry Point, where we were to be filming that morning – and later in the evening for the live show. The crew arrived the evening before so things were getting underway when I pulled up in my car as the sun was starting to rise. We were fortunate that the tides worked in pretty well and that the cameraman that I was with didn't have to wait long before a dorsal fin or two started to show up so that some footage could be obtained and 'in the can' to edit into the live show that would go out later that evening. One of the silhouette-type photos in this

book is of a member of the technical crew jogging down the beach with Kesslet surfacing only a few feet away from him. We took this as a good omen.

Filming during the morning went pretty much as I had thought with some dolphins hunting for salmon, some socialising, some doing a bit of bow-riding with a boat and everyone was fairly happy. Later we had to start thinking about the live filming in the evening and I was praying quietly that the available light and weather would hold good as the tide was going to be just on the rise as we went live to air at 8.00pm – would my dolphins turn up in time so that Simon could say, live from the Chanonry location something like "Yes, we actually have dolphins in the area and they are just coming into view now"? We had lunch on location and thankfully the BBC had supplied chemical toilets and other mod cons for the crew so we could stay put and not worry about such logistics. I was being bombarded with questions by the huge crowd of extra visitors (word that *Springwatch* is in town filming spreads quickly) about what was happening and who was the 'star' of the show – stuff like that. Time passed extremely quickly, and before I knew it we were being given our dinner before filming started again. The clouds were getting thicker and the tide seemed to be taking an absolute age to start changing, but my earlier prayers were

answered as Kesslet and her young calf Charlie plus Trail Scoop, Nevis and a few others started pottering around in the current near the Point itself. Through the rapidly failing light as the programme was getting ready to wind up, one of the dolphins caught a lovely fish and the cameraman that I was with got the whole thing on film, live as it happened. Looking back up to the car park we could see some of the production crew doing a little celebratory dance – job done, dolphins filmed, no technical issues – perfect. It was just before midnight before everything was tidied up and put away. It was a long day indeed, nearly eighteen hours on my feet, but a happy and successful one.

The next morning, Simon came and joined me down at the Point and we had the beach to ourselves. Dolphins came rolling in as if to say hello and we watched them for a while, took a few photos, chatted and generally 'decompressed' after a busy, full-on day before. Turning around after an hour or so of being transfixed by the dolphins, we realised that there were about a hundred people standing behind us – the *Springwatch* effect had kicked in with viewers seeing Chanonry on the telly and wanting to come and see for themselves which was great news for the local shops and businesses for a long time to come – *Springwatch* had really put the Black Isle, and the dolphins, firmly on the map.

Opposite: Simon King OBE at Chanonry Point on the morning of filming BBC's *Springwatch*, June 2008.
Above: ID#433 *Kesslet* surfacing only a few metres from a BBC crew member jogging down the beach, Chanonry Point.

Chapter 10

Things can go wrong

Sometimes things do not go to plan. This is, after all, wildlife filming and photography and the word to remember here is 'wild' – something that certain people in the profession should remember. Well-known presenters that I have worked with in the past like Michaela Strachan, Kate Humble, Bill Oddie, Simon King, Chris Packham, Miranda Krestovnikoff *et al* take the non-appearance for the cameras of completely free-ranging wildlife as part of the job. They will come to wild and very remote locations with a sense of hope, adventure and a positive attitude and breathe a huge sigh of relief when things turn out well. On the other hand, it's amazing just how unprofessional some people in the media can be when things do not go the way that they hoped. Here's a little 'for instance'.

A few years ago, through my esteemed WDC press relations manager and colleague Danny, I was made aware that a European film crew was desperate to do some filming with our dolphins and that they had requested that 'the dolphin man', as I seemed to be known across the English Channel, be made available to assist them. No problem says I (as usual) – and started a lengthy e-mail conversation with the director/producer trying to pin down what sort of footage they were after. It all sounded fairly standard but they wanted to come over in mid-February to do the filming – not the ideal time of year at all for either good weather or for reliable dolphin sightings. I tried a bit of friendly expert advice and suggested at least a month or two later in the season. No, came the reply, they couldn't come any later in the year and would have to take their chances – everything would be fine. Normally with film crews, I strongly advise that we try some filming from land at Chanonry for at least two, six-hour rising tide sequences. Then as a backup plan and for a bit of variety we do a full day's filming out on the water by chartering one of the local tour boats complete with knowledgeable and friendly skipper. This generally does the trick – everyone ends up very happy with lots of nice mixed land/boat/dolphin footage for the film crew, some more nice dolphin images for me and WDC, some extra much-needed cash in the pocket of the boat operator.

Against all my advice, the crew *eventually* turned up at Chanonry car park in a hire car from Inverness Airport looking tired, dishevelled and not giving off an air of professionalism at all. I told them as diplomatically as I could that, as I had warned them countless times before they even stepped on a plane, it was going to be a miracle if anything with a dorsal fin turned up to be filmed at all, that it was far too early in the season for the backup boat trip to be available and that the weather was getting more ominous by the minute. We spent a miserable three-and-a-half hours at Chanonry Point getting wetter and wetter (even in full waterproof gear) and a lot colder. We were also much too late in the tidal sequence as they hadn't arrived on time and of course we saw absolutely nothing. The next incoming tide was now going to be too dark for filming because of the dire weather, so I suggested that we try again the next day at the correct time and keep our fingers crossed. "Tomorrow?!" the director erupted… "We will be going back home tomorrow, no more chance for film!" With total disbelief in my voice I asked "Are you serious – you have come all this way for just one day, just one chance of filming completely wild dolphins at a bad time of year?"

Two months later another production company from a different country came over to do much the same type of thing and, as a complete polar opposite to the earlier debacle, we had an absolute ball. The weather was great, the dolphins were brilliant, the film footage from both land and boat looked fabulous and the film crew was also totally knocked out by the stunning scenery and the friendliness and hospitality of Scotland and her people. It couldn't have gone any better and to this day we still keep in touch.

Chapter 11

Not quite to plan

Staying on the subject of things not going *quite* to plan: a while ago the BBC arranged with Gwyn from Dolphin Trips Avoch and me to go out on the water to try and get some dolphin footage and do a wee report on them for the children's *Newsround* programme. It was a bit early in the season but we thought we might get lucky so we met at Avoch harbour and got kitted up and headed out to sea, after doing a bit of filming in the harbour. The weather was a bit rough and we hoped that it wouldn't get any worse as we headed for Chanonry Point, intending to go past the Point and over to the more sheltered coastline at Rosemarkie, hopefully picking up some dolphins on the way. The BBC cameraman was fiddling with some of the switches of his brand new, big broadcast camera. I was swapping over a lens from a wide angle that I had been using to take pictures of the presenter Leah Boleto to a bigger zoom in case we came across dolphins. As we came round Chanonry and started to feel the effect of the tide and wind together, Gwyn shouted "hang on tight folks" but just then a huge wall of spray hit me, the BBC cameraman, the sound engineer and all the camera and video gear including my open camera bag that had a Canon 7D camera body lying there with no lens attached to it. The bag was now full to the brim with salt water and the camera and the smaller wide-angle zoom lens were both drowned. The 7D camera body was literally smoking as I pulled it out of the bag and the cameraman had hastily pulled off the big battery pack from his £30,000 machine, hoping that nothing electronic had been fried. He was very lucky as nothing seemed to have been harmed – but one camera and one lens of mine was later written off by my insurance company (and replaced very quickly by them) and everything else needed to be rinsed in fresh water then dried before it could be used. We saw two dolphins for about 90 seconds. It still looked good on the telly though!

Chapter 12

It's all about the light

Late afternoon and early evening tides in the Inner Firth can give lovely light conditions for watching and photographing dolphins. As long as there are no clouds in the sky to make the water go too dark for filming or photographing you can technically snap away at dolphins in early summer right up to 9.30 or 10.00pm, as it doesn't get totally dark at all this far north at that time of year. I have stayed on at Chanonry Point many a time over the years after the cameras have been switched off due to lack of light and sat on the beach with friends and visitors watching these wonderful big predators still working hard and playing hard as the light fades away to what we call the 'gloaming', dusk as you might otherwise say. Sometimes if a kindly soul had driven up to Jaki and Sparks' chip shop in Fortrose for supplies, you could even relax with a hot bag of chips to hand – now that is civilised dolphin watching for you!

Late one sunny, summer afternoon I arrived at Chanonry car park and by pure chance drove straight into my usual spot – very lucky considering the time of year and, as usual for the summer high season, the car park was absolutely full. As I was gathering up my gear and just before I started to head down the beach, I had a quick natter with David Galloway who lives at Chanonry with his wife Mary. The two of them are not only lovely people but also birding experts. They scan the Point and beyond with scopes most days to see what bird life is around and let me know if they have noticed anything with a dorsal fin, or indeed anything out of the ordinary. David said that there hadn't been many dolphins around for the last couple of days but that he had spotted some Gannets further out, so maybe there were mackerel on the go. I was then chatting on the move to a few of the regular holiday dolphin watchers, some of whom had been at the Point since early morning and hadn't seen so much as a distant dorsal fin, nor very little the day before.

Sometimes during the summer you get the odd day when dolphins arrive in the current but don't detect any salmon, so after a while will amble off into the distance, but this was slightly worrying. I was meeting a producer and sound recordist from a popular radio station to do an on-the-beach interview about my work for WDC and to describe what the dolphins were doing. This would be a little difficult if there were a lack of dolphins and I smiled to some of the regular watchers that were waving to me. Some were saying things out loud like "Ah, it's all right – Charlie's here now, the dolphins will be along soon…"

Some people have the idea that I am like an aquatic Pied Piper and that dolphins simply turn up wherever I go – a lovely theory and certainly one that I wish was based on fact, but nothing could be further from the truth. Half-an-hour or so later the radio interview people crunched their way towards me and introduced themselves and we had a good chat about what we could get done that would be factual, interesting and hopefully entertaining too. I explained about the lack of dolphins just at the present moment and as I was trying to sound confident and reassuring, I have rarely been so relieved as I spotted some huge splashes away up the Eathie coast. Through my big lens I could see some dolphins travelling fast down the shipping lane and thankfully heading towards us. More dolphins were coming down past Fort George and from seeing nothing a half-hour previously, the water now seemed to be filled with dorsal fins as the two groups were beginning to coalesce.

However, none of the dolphins took up their usual positions to hunt for salmon in the tidal current just off the beach. Something else seemed to be unfolding in front of us and I realised what was happening and why no dolphins were seen earlier that day or the day before: big shoals of mackerel were heading in from the

open sea and the dolphins were following this highly mobile prey species closely and, working as an expertly co-ordinated team, were gradually taking turns at coming up underneath the tight compression of fish as the other team members kept circling the shoals. Although I had suggested to the radio interview crew that we might see some salmon being hunted just a few metres off the beach, I pointed out to them that this co-operative behaviour was a much more rarely observed hunting strategy. Although there wasn't the spectacular launching of big salmon and sea trout into the air to be seen, what was happening just below the surface was just as fascinating. By now some beautiful gannets were being attracted further inland by the prospect of a fishy dinner and these spectacular birds were wheeling around a few hundred metres above the groups of dolphins, that were by now heading further over towards Inverness Airport. Then the gannets started to plunge-dive. This thrilled the producer as she was an avid

birdwatcher and seabird enthusiast. Again and again these white missiles hit the water like arrows on an ancient battlefield, many of the birds striking their targets and I could see them bobbing to the surface swallowing the fish. I let the producer look through my 600mm lens at this, one of nature's great spectacles. The dolphins, still coming in from the open sea, were continuing to herd the fish successfully into manageably sized shoals and were steadily heading in towards the Kessock Bridge, around ten kilometres away. The radio interview went really well but I knew that very little was going to happen here now and suggested that we quickly get ourselves down to North Kessock and watch from under the Kessock Bridge before we ran out of light to see. We parked near the RNLI Lifeboat Station and watched the dolphins and gannets until it went dark. By this time the microphones and cameras were switched off and we just enjoyed the spectacle. What a great day, a late finish and the radio interview turned out not that bad either.

Page 39: ID#818 *Yin-Yang* travelling in the distance past Chanonry Point in a very glittery sea and glarey blue light.

Above: ID#744 *Squat Fin* aka *Bonnie* and tiny baby surfacing near Chanonry Point.

ID#969 *Spirit* and calf plus ID#568 *Chewbacca* travelling in very dark water with backlighting, Cromarty, Black Isle. 41

Opposite: two adult dolphin-and-calf pairs exiting the Kessock Channel in blue sea with the Chanonry peninsula and lighthouse 11km in the distance.

Above: ID#1025 *Charlie* breaching from the water at Chanonry Point.

Chapter 13

Flying visits

It is around ten o'clock on a dry, improving summer morning and the sea's grey tint is gradually turning to blue as the tide starts to turn around and slowly begins to rise. Dolphins are still away out in the distance with some over towards the airport shoreline and a few on the outermost reaches of Rosemarkie Bay. A motley selection of those who I call the 'village' are down with camera gear this morning – the dedicated and hardy bunch of seasonal holiday dolphin watchers and photography enthusiasts that take over the whole of Chanonry Point for most of the day. Many have become good friends of mine and of each other after coming back year after year after year. Pat and Brian, Jane, Mandy and Geoff, Jim and Dawn, Keith and Julie, Harry, Morag and Jim, Susan and Peter… the list goes on.

It is not long before some of the more local dolphin watchers start appearing: Big Al, Catherine and Karen, Ali, Alan, Francis, Paul – all kitted up with good camera gear, mixing with the 'village' with hugs and kisses going all over the place. All are ready to have a good session at the Point with the sound of cheeky banter and gales of laughter already ringing up from the shingle. It is not long before the first set of dolphins starts to appear from the south-east. It is Zephyr and her young male calf Breeze, who is having a mad five minutes breaching round and round his mum. They are closely followed by Trail Scoop (or Scoopy as he is sometimes known) – a lovely big guy always ready for a bit of hilarity and leaping around. From the other direction come Sundance and young Flake, one

of the sub-adult lads learning the tricks of the trade from a fully mature adult like Sundance. They meet up with the other dolphins around seventy metres out from the shore; have a chase around for half-an-hour with only Zephyr launching one good-sized salmon out of the water, then they all head off in the direction of Eathie almost as one entity, Zephyr travelling and swallowing the salmon at the same time as they disappear from view. That was it – short and sweet, forty-three minutes from start to finish is all that the dolphins would give us today. Who knows, tomorrow it could be in excess of six hours. That's wildlife for you.

The next day in fact wasn't quite six hours' worth of dolphins but not that far short; and what they lacked in hours of attendance they were certainly making up for in physical and social activity. It was a bit grey, chilly and rough compared to the day before and I was having a bit of trouble with one of my cameras, my backup Canon 7D, the autofocus system of which was suddenly misbehaving. It wouldn't stay locked on a subject for any length of time and my main camera was away at CPS Elstree being repaired after falling from a great height, bending the lens coupling on the camera body but miraculously not harming the lens at all. I was getting worried that my day's work was coming to a premature end. But, when lifting the mirror of the 7D to have a look at the autofocus array underneath (many people don't know about this critical part of the camera), I found that there was a great wodge of fluff covering the

Above: the tail of ID#433 *Kesslet*.
Opposite: ID#997 *Comet Too* breaching from the water near Chanonry Point, Black Isle.

middle AF points – no wonder the poor camera couldn't 'see' what it was meant to focus on. Goodness knows where this fluff had come from but after being cleaned out, much to my relief it worked better than it had done for ages and I got back to the dolphins. Mischief was having an absolute ding-dong confrontation with Rainbow's big son Prism: they were head-butting each other, biting each other's tail stocks, rolling over each other and breaching. It was brilliant to watch but wasn't serious fighting, more a matter of robust play as there was no blood being spilled anywhere. After twenty minutes or so they calmed down and got back to the important job of the day – getting some fish.

There were a few boats in the area being a nuisance to some dolphins who were trying to forage over nearer the Ardersier coastline, but after a phone call to Gwyn Tanner from Dolphin Trips Avoch (who was out on the water on a trip with passengers) to go and have a polite word with them, the boats thankfully left the area, letting the dolphins get back to hunting. I noticed out of the corner of my eye the small shape of a young Harbour porpoise sneaking out of the Chanonry Narrows in between the groups of dolphins and he or she made it through the dolphin 'minefield' unscathed and out to the open sea. A lucky escape as our dolphins will often attack porpoises, sometimes killing them. Closer to shore however, things were hotting up a bit with a large group of females and calves of various ages invading the area around Chanonry. Some of the big females like Tall Fin, Moonlight, Happy Dragon, Porridge and old Jigsaw were each grabbing some decent-sized fish and scoffing them down before heading off into the distance. I was by now having a very good day, Photo ID wise, as I had good, close-up and sharp pictures of Moonlight, Rainbow, Sundance and

Mischief, four out of the six adoption dolphins that I look out for and study for WDC. Some of the dolphins that were still foraging around Chanonry Point obviously felt that the crowds of people watching from the shore were not nearly excited enough, so a big game of 'chases' started. Rainbow and Squat Fin joined Mischief, Zephyr, Prism and Sundance plus a few others and soon there were dolphins whizzing all over the place, making lots of spray and half breaching out of the water – great to watch but terrible to try and photograph. Time, therefore, to put the camera and big lens down, sit and have a drink to relax and rehydrate and grab a bite of lunch.

The other dolphins that were around I counted as a real bonus, especially the 'Cromarty gang' who are dolphins that are more often spotted by Barbara from the Lighthouse Field Station out near Cromarty and further north while out on boat surveys, or by Sarah when out on the water with customers on Ecoventures. It is always lovely to see a big group of mums and youngsters rolling along together and today I was blessed with some additional and not-so-familiar dorsal fins, the identities of which I would have to double check later on… variety is the spice of life, as the saying goes. After a busy day when I might take well over fifteen hundred photos, I often find dolphins that have been masked or hidden behind others when I am going through the images back at our Studio in North Kessock and have to re-estimate that day's number of individual dolphins encountered. I think that my record for the number of good-quality frames of dorsal fins taken in the space of one day at Chanonry and dolphins identified from them is in the region of fifty – just to put it in perspective, that is a quarter of the estimated population. Not bad for a day's work.

Chapter 14

Generous friends

My pal Alan Ward and I are standing having a natter (nothing new there then!) outside the Ecoventures office in Cromarty on a dull but rapidly brightening June morning, waiting for the others in our group to arrive. We are here pretty early as we have been invited out yet again on a privately chartered boat trip by our friends Andy and Elaine White, lovely generous (to a fault) people and great WDC supporters. They prefer to have Sarah's lovely boat *Saorsa* (Gaelic for freedom) pretty much to themselves and tend to charter the boat early in the morning before the normal daily trips start. These early jaunts can be interesting, not only from a perspective of light for photography but for the lovely peaceful atmosphere as you put out to sea.

Before we know it we are all here, have lifejackets on and Sarah is guiding us out towards the Sutors and we head over northwards towards the bird colony where we realise that we are not the only early risers today. The team from the Lighthouse Field Station are out on an early dolphin Photo ID trip and have dolphins with them, so after taking a few distant shots of a dolphin half breaching beside the *Rona* we head out just beyond the Cromarty Firth entrance and pretty quickly pick up some dolphins of our own. Young Kenobi, big calf of Chewbacca, is pottering around catching small fish. Not far from her and coming towards us quite quickly are Rainbow and her calf. This is lovely as I haven't seen much of Rainbow all season and her calf, now over two years old, is looking in good health and both of them are swimming beside us at a leisurely pace. Then, just for a change, they dive underneath us and come up the other side as if lining up to get photographed. Around fifty metres away is another very distinctive dolphin, Jigsaw, one of our very mature females who has an almost luminous white

fringe on her jagged dorsal fin making her easy to spot from a great distance. Like Kenobi, Jigsaw is hunting for fish too, so after a few photos are taken we leave her and head out towards Nairn where the water is lovely and calm with just a tinge of blue on the surface. Big splashes in the distance herald the presence of more dolphins, which are making a bee-line for the boat and closing in rapidly.

Sarah slowly and carefully turns the boat around as we need to start heading back to Cromarty soon anyway. Sure enough, we end up with a big group of dolphins scattered all around us, keeping us company beautifully. Some dolphins are right underneath us but some others, young sub-adult males, are showing off – breaching about twenty metres away and showering everything with spray. These characters are some of the *Bad Boys' Club* as I call them, and the name says it all – if there is any mischief to get up to or trouble to get into, these guys are right in the middle of it and they are truly wonderful to watch. It's young Flake and his buddy Yin-Yang who are breaching a perfect distance from the boat for me to fit them all into the shots and the light conditions are lovely. Coming up beside us on the other side of the boat is Midnick and a few others of the girls including Spirit and her younger calf Shimmer, so I am well pleased at getting nice new shots of Rainbow earlier and now Spirit, two of the dolphins that you can adopt with WDC. The dolphins pretty much escort us back to the Sutors cliffs near Cromarty and as time is marching on we have to wave a fond farewell to them and make our way back to Cromarty Harbour so that Sarah can start her scheduled day trips for the public. Another cracking trip with the Whites completed and Alan and I are still grinning from ear to ear as we say our goodbyes and head away from Cromarty… until next time.

Opposite top: ID#815 *Prism*, son of *Rainbow*, breaching away in the distance, Inner Moray Firth.
Bottom: three of the *Bad Boys' Club* with the South Sutor coastline near Cromarty in the background.

Chapter 15

Chanonry days

It is a lovely early summer afternoon at Chanonry and the tides are rising about 3.00pm. There are blue seas and blue skies with just a hint of cloud and I am slotting my new lightweight 500mm lens with the 1DX camera body attached onto my lately acquired heavy duty carbon fibre gimbal head for the big Gitzo tripod. Alister, an old buddy of mine and also the manager at Ffordes of Beauly camera shop, had persuaded me to try it and I was well pleased with it – the very smooth fluid-damped action is great for precise control, especially when the wind starts to pick up and these

big lenses with huge lens hoods become like airport windsocks, blowing and vibrating all over the place even with the magic of built-in gyro image stabilising. I am in my favourite elevated position back up the shingle hill nearer to the picnic tables than the shore as I'm primarily using my big fixed-length lens today. There are some dolphins away over towards Eathie and they are progressing slowly down the coast with a bit of breaching going on to whet the appetite of the gathered watchers and photographers. It has been ages since we had a good bit of action at Chanonry so I'm

keeping my fingers and toes firmly crossed. One of the tour boats has just gone past and there is a couple in sea kayaks hanging around hoping for dolphins to paddle with. A few dolphins including Zephyr, Breeze plus Bonnie and her calf arrive in the tidal current that has only just started to move. The dolphins hug the coastline so closely when making this approach to Chanonry Point that they often take you by surprise. The arrival of these dolphins starts the rush to the water's edge by visitors and some of the enthusiast photographers. It is still a nice afternoon… couldn't really be better for photography; the water is now a very deep blue and I'm watching the clouds thickening a bit in the sky directly behind me as they are making the water darker all the time.

From the direction of Fort George, about a kilometre across the water from Chanonry Point, we see a series of huge splashes in the water and it is apparent that there are at least two big adult dolphins interacting with each other. They are coming in the direction of the Point fairly quickly, both big dolphins moving rapidly and breaching out of the water together. By this time I'm desperately trying simply to fit them into the camera's viewfinder – the one disadvantage of having a fixed focal length lens is that as the subject approaches you and gets bigger and bigger in the frame there really isn't much you can do about it. Thankfully both dolphins – who I now recognise as Rainbow's lovely big son, Prism and Trail Scoop (another of the big local males) decide to continue with their 'I can jump higher than you can' contest in pretty much the same place and distance from my position that I can just keep both of them within the frame, but it is very tight. They breach out of the water together five or six times in succession, Prism eventually getting the gold medal for the highest breach – about five metres in the air and horizontally over the top of Trail Scoop at one point – much to the awe and delight of the gathered holidaymakers who are screaming and clapping. I am concentrating pretty well

for a change and only miss one double-jump sequence amongst this lot of multiple breaches because of the shingle slope suddenly giving way below my feet, a potential hazard whenever you have to quickly relocate your strategically chosen position, and all the camera gear, to get away from people who unwittingly stand or wander right in front of you!

Things quieten down just a little as the other group of dolphins that I spotted earlier begin to arrive in the tide from the Eathie direction. I start to pick off and photograph dorsal fins of the dolphins that are coming into my viewfinder, namely Mischief, Sundance, Mini Fin, Flake and Bodhi. They all seem to be males, adults and some sub-adults. This could be very interesting as we already have two male dolphins nearby that are very much filled with testosterone. Encounters between groups of roving lads can end up getting a bit lively and sometimes violent, but thankfully, catching a late lunch takes over as the main job of the afternoon with all the dolphins settling down to a bit of more or less peaceful hunting. Some break away from the Chanonry group after half-an-hour or so and head down in the direction of Inverness. I visually follow them by looking through my big lens, now with a 1.4 magnification converter fitted taking it up to 700mm in focal length. A few females and calves join them now as they head south into the blue haze, with the occasional jump by one of the younger males being thrown in just for good measure. It is still a nice afternoon and there is plenty of tide still to come in but it seems that the salmon run just wasn't very good or sustained so, although some dolphins are still pottering about in the tide for a while, all the exciting action has pretty much finished for the day. I comment to a photographer buddy, Andy Howard, who I had noticed arriving on the beach just as the action was starting, "You jammy so and so – there has been nothing happening here for weeks then you turn up and it's like a damned circus". Talk about being in the right place at the right time!

It is a wild and dull afternoon with a north-easterly wind keeping the temperature way down, despite it being June and the sea state is very rough with a big swell. There are some small boats out and about but most are running for shelter including some lovely veteran dinghies that are taking part in a historic boat festival. One of them comes bouncing past Chanonry Point heading for Fortrose Bay and even thinking about it now as I type this story makes me feel seasick. Dolphins are starting to patrol around as the tide is on the rise and I have already spotted Sundance very close to the shore; just beyond him is a lovely female called Chewbacca who has her young calf Yoda with her. It is difficult to get an autofocus lock on their dorsal fins as the waves and spray keep masking them, but with a bit of fiddling around with some camera settings I manage to settle down and get some sharper pictures. A larger, taller and very tattered dorsal fin goes tearing past me – the dolphin already locked onto a salmon. It is big Nevis who has arrived from nowhere and he is sending spray everywhere… always the showman. Through the waves comes another female with a calf

Above: ID#815 *Prism* with Chanonry Ness Lighthouse in the background.
Opposite: ID#105 *Sundance* breaching towards ID#815 *Prism* at Chanonry Point with Ardersier village in the background.

in tow but I can't see who they are at first as they have dived quite deeply – I get them in my viewfinder as they surface in unison and it turns out to be Sickle and her cute two-year-old. By this time things are getting a bit on the hazy side and turning the big Canon 400mm F2.8 IS lens round I can see that the huge front glass element is covered in sea spray. The problem with standing at Chanonry in a strong north-easterly wind is that it blows right into you; the salt also makes your eyes sting so after a few hours working in these conditions you have some very salty camera gear and a pair of red, stinging eyes. The best thing to do is to go to your rucksack, grab a bottle of water and rinse your eyes so at least you can see again, and basically do the same for the front of the lens. Trying to wipe off salty spray with a lens cloth just makes a cloudy, greasy mess so the judicious application of some fresh water takes away the salty film and makes the lens much easier to clear properly. My big pro lenses are weather resistant so I can liberally apply some clean water to wash off the corrosive salt. But please be careful with your own camera gear no matter what level: use a damp cloth and wipe – but better not pour over your lens like I tend to do.

Anyway, back to the dolphins… one or two more individuals are arriving from the Rosemarkie Bay area and joining in the salmon hunt and I can see some big fish being tossed in the air through the spray. Zephyr and her young son Breeze are very near the beach in the choppy water and she is regurgitating a big fish. Moonlight appears for a brief visit but heads back out towards Fort George along with Sundance, Chewbacca and her calf and Trail Scoop who has appeared from nowhere. Nevis is anchored to his favourite spot, right at the convergence of two swirling currents about sixty metres offshore and all of a sudden he is the only dolphin left near Chanonry. I sit down on my seat and rinse my eyes again, have a drink and try to get back up again but my legs and body are saying 'no way' as I have been fighting the wind and spray for about five hours now. Soon even big Nevis ambles off into the distance as the tide is now much higher. That finishes me for the day. Then the wind drops and the sun comes out.

ID#866 *Zephyr* in typically vertical mode with ID#748 *Trail Scoop*
aka *Scoopy* breaching horizontally behind her, Chanonry Point.
Page 58: ID#672 *Munchies* breaching from the water on a lovely
sunny day at Chanonry Point.

Chapter 16

Rainy days and Mondays

Dull and rainy days are not normally ideal for obtaining nice, colourful pictures of dolphins but there is a slight advantage sometimes in that the lack of sun glare can let you see the actual accurate colour and hue of the dolphins' skin. The real colours are a dark slate-grey back graduating to light grey-white along the sides and then pure white underneath – classic oceanic countershading, the same as in many other species of marine life, like sharks for instance. In bright sunshine, the skin of our dolphins takes on a tan brown colouration as the ultraviolet rays get diffracted and blurred by the water and then hit the melanin of the skin. Our dolphins don't have perfect, blemish-free skin by any means; they have skin conditions, rashes, cuts and scrapes, just as we do. Sometimes there are coloured patches in certain areas that look very different from the surrounding epidermal layer – these are called skin lesions of which there are quite a few different types. One of the most frequently seen but little understood types are the

Page 59 top: at Chanonry Point, ID#23 *Mischief* and ID#744 *Squat Fin* aka *Bonnie* passing people standing on the shingle beach in late afternoon sunshine.
Page 59 bottom: dolphin watchers on a sunny but very windy evening, a lovely blue sea but not warm, even in July!
Above: ID#969 *Spirit* and ID#1042 *Legion* in very monochrome light, Inner Moray Firth.
Opposite: ID#1020 *Idris* giving a little breach in misty conditions, Cromarty Firth.

mustard-yellow patches, almost like lichen or algae that appear on young dolphin calves' dorsal fins, faces and sometimes the rest of the body too with varying degrees of severity. It appears and then seems to fade away as the calf grows up, rarely seen after about three years of age. It may also affect adults but as their skin is much darker in colour maybe we just can't see it. This and other lesion types like depigmentation around the edges of fins and pitted, often painful-looking blemishes could be something that dolphins put up with all their lives that don't really affect their health to any great degree and are just part and parcel of living in an oceanic environment. Even stranger is the fact that we notice years where some dolphins seem to be much less affected and yet in other seasons they can look particularly marked.

Now, I'm not advocating that you purposely stand out in the pouring rain to take photos of dolphins, but over the years I have taken many, many frames in grey, monochrome light and sometimes you do get some very interesting and arty photos. Remember that dolphins are primarily grey on most of their bodies, so mix that with the surrounding grey water and you can get some interesting blending of tones. I remember a rather grey and mystical foggy trip with Sarah at Ecoventures. Despite a variable sea haar, we edged our way out of the harbour towards the South Sutor area of the Cromarty Firth and were quickly joined by some dolphins that spookily appeared and disappeared again out of the rapidly thinning fog. The horizon, sea and sky were all the same tone and shade and some of the photos from that day came out

Above: a group of sub-adult dolphins passing very close to the beach at Chanonry Point on a grey, rainy morning.
Opposite: ID#1025 *Charlie* son of ID#433 *Kesslet* breaching so close to the camera that the author cannot fit the whole dolphin in the frame. Very flat, grey light.

as though I had painted them in watercolours using only the one paint. Fifty shades of grey? More like a hundred and fifty. When it is raining around the Inner Firth area the water often tends to be quite calm and photographing dolphins with water droplets from the rain all around them can be quite pretty, adding to the spray made by the dolphins breathing and splashing as they surface and dive. Dolphins are very tactile animals and in very heavy rain I have noticed one or two of them lingering at the surface for a while – maybe the rain tickles them and so they lie on the surface for a bit longer than normal to make the most of the sensation.

Page 64 top: ID#433 *Kesslet* looking almost like a calf compared to an (unidentifiable) adult male in very backlit conditions. Bottom left: an adult dolphin breaching in the distance on a misty day, Cromarty Firth. Bottom right: two young dolphins silhouetted in backlit conditions – terrible light to identify individuals that do not have big nicks out of the dorsal fin.
Above: ID#192 *Arabic* lifting her tail out of the water as she deep-dives, giving a lovely cascade. Left: ID#989 *Spirit* swimming through very calm, monochrome water.

The scarred head of ID#23 *Mischief* appearing from the dark, mirror-calm water giving a lovely reflection.

Chapter 17

I can't feel my nose

The winter months can be a tedious slog just trying to find dolphins at all, never mind actually photographing them, but sometimes you get lucky, especially if there are some of the more local dolphins still around in the Inner Firth that are finding some of the seasonal menu to keep them going over the lean months, prey species like mackerel and sprats. In Fortrose Bay, over to the right of the Chanonry Point car park, there are two red Northern Lighthouse Board navigation markers, identifying the Skate Bank sand bar and through the grey, dull light I could make out a bit of splashing going on. Sure enough, there were three or four dolphins playing 'Follow my Leader' around the nearest buoy and as it had just stopped raining I decided that it might be worth getting out the tripod and camera with my big lens. By this time my buddy Alan Ward who lives in Avoch – a village only a couple of kilometres away – had arrived. I had texted him right away on seeing dolphins and now he was watching what was going on and giving me a running commentary on what they were doing. This was extremely handy as it meant that I didn't have to search for them again after I'd set up the camera gear. Kesslet and young Charlie were having a great game with Bonnie and Trail Scoop, zooming around the red buoys and then ambushing them as they came back round again. It was great fun to watch even though the poor light wasn't going to help the photo quality but beggars can't be choosers – and I was just thankful that we had managed to see some dolphins at this often bleak time of the year. I managed to get a few grey but at least sharp photos that would make a nice blog entry and updates for a few social media sites like Facebook and Twitter for the first time in ages.

Winter can also be a time up here when the weather is your biggest and most constant enemy, snatching away any available visibility and replacing it with sleet, rain and thick cloud. Every so often however, if there is a really long cold spell then the sea flattens out due to high atmospheric pressure and the air is much crisper and clearer – a lot better for long range observations through the big, crystal-clear Minox 15 x 56 binoculars that have been loaned to me by Minox UK. For really distant observation, I can attach a nifty little eyepiece that fits onto the coupling of the telephoto lens which turns it into a monster telescope. Many an hour I spend during the winter and early spring, looking out over the Inner Firth, either sitting in the car or standing on the grass of the Rosemarkie Caravan Site while it's closed for the winter. It provides a very handy raised platform as it gives a great view of the coastline all the way up to the South Sutor cliff, many kilometres to the east, and right across the sea past the seal colony at Whiteness Head, down to Fort George and over the narrows to Chanonry Point itself. A great number of overwintering birds arrive here – species like Long-Tailed ducks, Wigeon, Teal, Goldeneye and Merganser. We have three species of Diver, Red and Black Throated and the lovely big Great Northern that often stop for a while in Rosemarkie Bay before flying on up to the lochans in amongst the high hills. Sometimes the Long-Tailed duck are a nuisance, as when they are in big groups and are chasing each other, the splashes that they make are very similar to distant dolphins and can really fool you.

We had a really bitterly cold spell up here in early February a few years ago and some winter herring started to arrive quite near Chanonry Point with incoming tides, which in turn brought a few groups of dolphins in for a couple of days to see what they could snap up. One day the weather was really fine, bitterly cold but bright and clear – I love days like these as the atmosphere is so clean and distortion-free – you can see literally for miles. I set up my seat/rucksack and lowered my big tripod down so that I could sit comfortably and use the camera with the big lens on the gimbal head and started watching. I had the place to myself – there was nobody around at all, nice to have 'my' beach back for a while. Over towards the cliffs below Hillockhead I saw some small splashes and a few plumes of dolphin breath, highly visible in this very cold air. The venting of the warm air from their lungs can often be seen before you spot the dolphin itself. I could make out around six or so individuals; one adult had a small calf with it and they were spread out in an even line as though waiting to start a race. All were facing into the incoming (very gentle) tide and were pointing away from my camera. I could take photos and then zoom in to them via the screen on the back of the camera but even so, they were at such a distance that I was struggling to identify any of them. Another, larger, dolphin group had appeared about five hundred metres further up the coast and was facing the group already there and were all sitting in a similar line. Nothing happened for a while; it was a bit puzzling for a minute or two but then both groups headed towards each other at high speed. They were really motoring and as the groups merged together I could see lots of small fish leaping for their lives out of the water, their pinkish-silver colouration glinting in the sun. The dolphins were using a hunting tactic that I had never witnessed before. I had watched them 'corralling' or rounding up a shoal of fish before many times but this was different, a variation on a theme.

It just goes to show – never get smug or complacent about wildlife, especially very intelligent species like dolphins and don't think that you know everything about them – they will sometimes do totally unexpected things that will leave you speechless. I watched this spectacle happening again and again, wishing that the dolphins would get a bit closer for the sake of useable photos but they, or rather the fish, seemed to be stuck in the one area. It wasn't just them that was stuck either – I tried getting up to move the seat and tripod back as the water was rising a bit faster now and was near my feet. I was pretty much frozen to the spot. I had been so engrossed in what the dolphins were doing that I hadn't realised just how cold I was getting. The back of the camera was iced up with the vapour from my breathing on it for so long and my legs had almost gone to sleep, thankfully not from frostbite but severely stiff with the cold. I got to my feet just in time and dragged everything back as the water started to lap at the bottom of the tripod legs. I decided that after four hours it was time to head back for the car and warm up a bit. A wee tip here: taking a freezing cold camera and lens into a warm room or a rapidly warming up car will mean that the whole thing will very likely 'fog' up with both internal and external condensation. If you leave it for long enough it will go away but it could take ages. The bigger the lens and camera, then the longer it will take. To reduce the condensation problem, put the camera/lens inside a big plastic bag and tie off the open end. The condensation will be transferred to the sides of the bag instead of inside the lens and camera. If you take a warm camera into a cold environment it's not quite so bad, but you should leave it for a while until the temperatures start to equalise. And remember that cold drains camera batteries very quickly. Keep a spare battery at body temperature inside your jacket and swap over when the battery warning light starts to flash.

As such, the weather doesn't worry me as you simply dress accordingly and professional-grade camera equipment is built to be weather-resistant. It's worth buying a pull-on, full-length and tailored-to-fit waterproof cover for the camera and lens, Kevin Keatley from Wildlife Watching Supplies does very good tough covers.

The biggest pest when you are trying to work in windy, wet weather is getting the front element of the lens covered in spatters of sleet or rain, and when it comes to the huge telephoto lenses that are used in my line of work, they can take a lot of wiping to get clean as I have already mentioned. Technology, however, is improving all the time, making this onerous task less of a chore these days as new, non-stick, anti-smear optical coatings are being applied to the front elements of newer lenses and the water just runs off. Good, eh?

Opposite: ID#1022 *Yoda* son of ID#568 *Chewbacca* doing a lovely twisting breach on a cold day in rough water at Chanonry Point.

Chapter 18

Out of the blue... into the green

I am sitting in my car early one morning at Cromarty, having arrived way too soon for the planned all-day photo identification research boat trip with my friends at the Lighthouse Field Station, so I'm watching the water for signs of life. Sure enough, two dolphins go ambling past close to the south shore heading out to sea, passing the two or three campervans that are parked on the grass near the Emigration monolith stone. I wonder to myself if any of the people in the vans have seen or heard them – what a way to be woken up… by dolphins. I take the sighting of the two dolphins as a good omen as Barbara, Becky and Sarah arrive at the Field Station and the rigid inflatable research boat *Rona* is wheeled out on her trailer, towed by the lovely old, but fully restored, red tractor.

After the boat is checked over and we head down to the slipway, launching goes smoothly as usual and before we know it we are all safely aboard. With both engines running smoothly we

are heading out steadily between the two land masses known as the Sutors towards the mouth of the Cromarty Firth – an area of very deep water that merges with the Moray Firth itself and a very picturesque area of coastline it is too. It takes us quite a while to find our first dolphins to photograph – we are nearly at Nairn as we come across a maternal group with some familiar dorsal fins. They include a lovely female called Tall Fin and two of her offspring, Bodhi, a strapping lad and his young sister Doyle who are rolling along smoothly through the calm blue sea. Just a little way from

them is Porridge and her two-year-old calf. Our first dolphin encounter of the day has begun. The speed of travel and surfacing pattern of the dolphins is perfect for getting identification shots and within a minute or two both Barbara, who is the main Photo ID photographer, and I have plenty of good, well-focussed frames.

We see some splashing away in the distance and more 'customers' start to come within camera range. We are both using 70-200mm F2.8 Canon zoom lenses which may sound big but when there is a lot of sea between you and the dolphins you always wish that you had more focal length. Pretty soon though, the lenses are more than large enough as a mostly female group comes up rapidly alongside us, with great favourites of ours including Midnick, Muddy, Moonlight (one of the WDC adoption dolphins) plus Sickle and her small calf well within camera range. Another of the adoption dolphins, Spirit, with her daughter Sparkle and her younger calf Shimmer are merging with this maternal group and I am very pleased at getting new, charismatic and nicely lit shots for WDC of Spirit and Moonlight and friends. It is such a lovely day, the sea and sky are pretty much the same colour, and the light quality for photos is sublime. Becky carefully manoeuvres the boat so that we are slightly behind but on the other side of this, by now, quite big cluster of dolphins. This means that we can move up carefully level with the group and photograph the individual dolphins from the other side so that we have both

Above: four members of the *Bad Boys' Club*, ID#990 *Flake*, ID#818 *Yin-Yang*, ID#815 *Prism* and ID#1018 *Bodhi* larking around with the Black Isle coastline in the background.
Opposite: ID#23 *Mischief* surfacing through mirror-calm blue sea creating an almost perfect reflection.
Pages 72-73: ID#1109 *Puddles*, offspring of ID#11 *Muddy*, seemingly pushing clouds along with his nose. The water is almost like liquid blue glass and the sky is reflected in it.

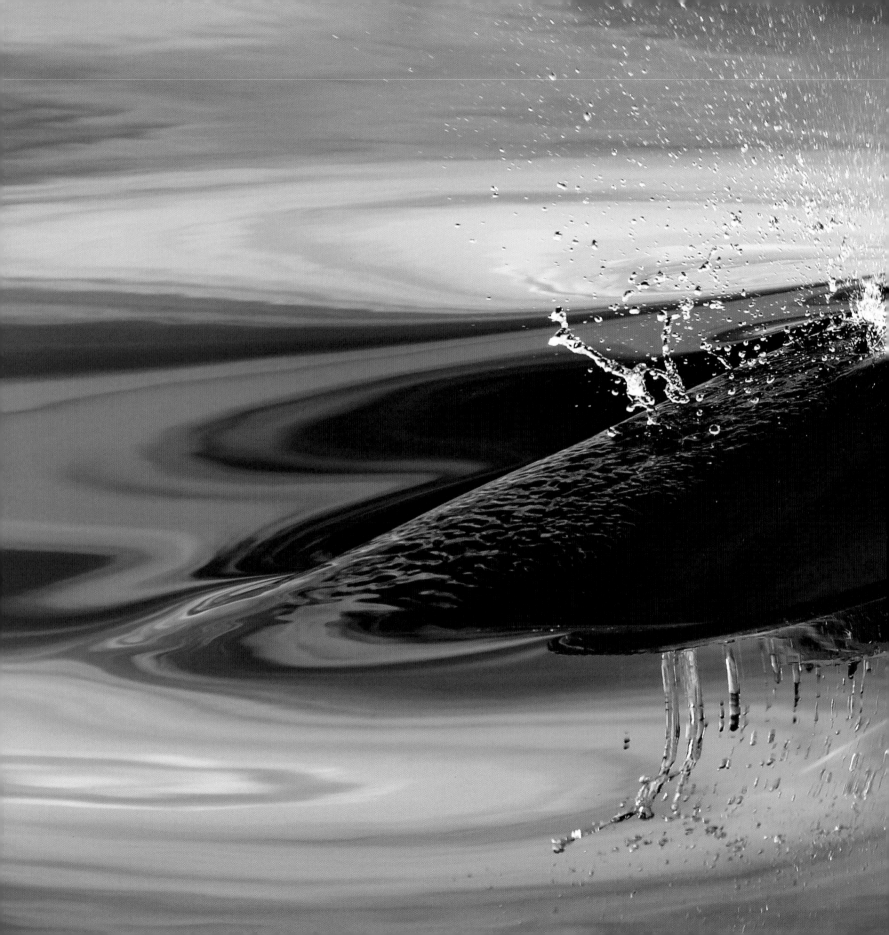

left and right side dorsal fin photographs. This is important as the dolphins can have different natural markings on either side of the fin. By this time we have tracked our way back towards the Black Isle coastline and we wave a distant hello to Sarah Pern from Ecoventures, who has a boat full of guests onboard *Saorsa* and they have quite a few dolphins for company too – and big smiles all round.

A few kilometres further on we decide to stop the boat, take a lunch break, and have a quick rest but after twenty minutes or so there are some very big splashes in the near distance and

very soon we have some dolphins arriving nearby who dictate an early halt to our lunch. There are some very big individuals appearing, some of the large adult males that we know well like Denoozydenzy, Yin-Yang, Raggedy Fin and Trail Scoop. They have the most amazing markings on their dorsal fins, generally caused by each other: bites, linear tooth-rake scratches, notches and nicks in the trailing edge – typical 'blokes' that spend half of their time proving themselves to each other. Sometimes these contests can get pretty violent and dolphins' pointy teeth can do a lot of damage – thankfully the skin heals over the scratches quickly but the chunks of tissue, or 'nicks' as we call them, that are bitten out

Opposite: ID#815 *Prism* breaching in a lovely calm blue sea with the Moray coastline in the background.
Above: ID#1109 *Puddles* looking up at the author through crystal-clear water with the surface tension buckling the reflection.

down the trailing edge of the dorsal fin are permanent markings. These are ideal for re-photographing at a later date, this research method being known as 'mark-recapture', which is very reliable and used on several different species of cetacean, sometimes using the trailing edge of the tail flukes of big whales like Humpback and Sperm. But I digress: our dolphins are trundling along at a reasonable pace but for once they are keeping a perfect distance from the boat – not too near, not too far, just right for getting their impressive dorsal fins squarely in the frame. Even better, they decide to fall back behind us and then pass us on the other side so we get shots of the fins from both left and right. Thanks boys, very co-operative of you!

The group of males tears off into the distance and then all hell breaks loose as they seem to encounter a run of fish that could be either sea trout or salmon. Some colossal explosions in the blue water are followed by fish being hurled many metres skywards, some very large dolphins then making sure that they are catching the fish when they land. We can see more dolphins in the distance which Barbara doesn't think we have photographed, so we head slowly towards them to see who they are. Amongst them is a big old male dolphin, ID#1 – Singers, who was the first Moray Firth individual to be catalogued by the researchers away back in 1989. He is a dolphin that I don't often see around Chanonry so it is a particular treat for me to spend some time with him. A few hundred metres away there's a small group of adults and calves and I am delighted to see that Rainbow, another of WDC's adoption dolphins, is there with her two-year-old calf. The two of them are quietly ambling along with some other mum-and-youngster combinations like Chewbacca and Kenobi plus Echo with her first calf, which is lovely to see. Keeping them company is the huge figure of Beatrice who, despite the female-sounding name (he was actually named after the Beatrice Oil Company) is one of the biggest of the male dolphins that we know of with an impressive 'passport photo' of a dorsal fin. This group is almost in resting mode, travelling along slowly and quietly in very close formation, so we decide to leave them in peace and head back towards the Cromarty Firth.

Above top: a young dolphin surfacing to breathe in mirror-calm water with clouds reflected. Bottom: ID#818 *Yin-Yang* travelling along though beautiful calm and clear blue sea. Opposite top: ID#1109 *Puddles* venting as he surfaces to breathe. This image won the author a national award. Bottom: ID#23 *Mischief* with water spray all round him as he surfaces right beside the author.

By this time the afternoon is stretching into early evening and the shadows are lengthening, but we still have some dolphins waiting for us to photograph as we come past the Buss Bank navigation buoy at the entrance to the Sutors. In the shadow of the North Sutor cliff there are two dolphins that seem to be chasing fish – an adult called Goose and a younger dolphin called Puddles. The light on the water is fantastic, turning the whole sea an amazing colour of emerald green. The way that Goose is exhaling leaves a plume of bright white vapour hanging in the air which is lovely to see. After a while these two are joined by Raggedy Fin who thinks that he should come and see what is happening. As we already have them all well photographed and, given the time, we agree to call it a day. After nearly ten hours on the water and around two-and-a-half thousand photos for me to look at later – quite a long day it has been.

It was one of these ID survey trips where all the factors came together for perfect dolphin photography: the fantastically blue water, glass-like tranquillity, crystal-clear light and the way the dolphins were surfacing right beside us was almost dream-like. I took one particularly rapid sequence of images of a young dolphin called Puddles coming to the surface to breathe. In one frame, in which he is venting out as he comes to the surface, with the surface tension still holding the liquid around his jaws like plastic shrink-wrapping – and with the water droplets hanging in the air like ice crystals, it looked pretty special. So, although I am not normally a competitive sort of photographer, I decided to enter it in my first ever national photography competition and it won. I became the proud owner of the title 'Scottish Nature Photographer of the Year'.

ID#963 *Goose* venting in green water, Cromarty Firth.

Chapter 19

Coastal adventures

A few years ago I was asked to assist with filming for a television series and the floating 'platform of opportunity' that we were going to be using for the whole day was the big converted lifeboat that operates wildlife trips out of Buckie Harbour – the *Gemini Explorer*. Husband-and-wife team Davey and Iris Smith who run the boat are old friends so I arrived early to catch up with their news and gossip. Soon after, we welcomed on board the camera crew and presenter, Hermione Cockburn, who worked on the BBC Coast series. Hermione and I had met briefly before while filming something else, but I wasn't going to be on camera much for this day's shoot so I could just enjoy myself helping out technically in the background and hope that the lovely-looking and warm day that was unfolding would be filled with great footage of dolphins and the beautiful Moray coastline. Heading out from Buckie harbour and travelling a wee bit east, we stopped for a quick look at the seal colony on Craigenroan Rock. Mostly, Grey seals haul out here, but there were a few Common or harbour seals today too. As we headed past Findochty, two dolphins, an adult and a juvenile, came over to us for a brief look before turning and heading away – at least we had seen two dolphins, albeit briefly.

The producer and I hit it off right away – a lovely person, she really wore her heart on her sleeve when it came to dolphins, jumping up and down with very moist eyes and squeaking with excitement at the sight of just a dorsal fin or two. I should have made sure that my camera bag had plenty of hankies in it for what came next. Just as we sailed past the village of Portknockie with its wonderful natural stone arch known as the Bow Fiddle Rock and entered Cullen Bay, around eighteen or so dolphins, some pretty well known to me like Black and Decker, Runny Paint and Neptune came alongside us on both sides of the boat and kept us company right across the expanse of the bay. There were three small calves with their mothers in particular that kept surfacing right beside us as if to say 'can we be on the telly?' The cameraman and sound technician were finding it difficult to know where to look and point cameras and microphones as there seemed to be dolphins simply everywhere by this time, some of whom were bow-riding too. Hermione was grinning from ear to ear and pointing out dolphins to the camera here, here, and here – oh, and over there! I was doing my own thing trying to get identification shots and the cameraman took my lead when he wasn't filming. Basically, anywhere I was pointing my camera and biggish lens (a 300mm F2.8 IS if I remember correctly) he did too and the grin on his face said it all: wall to wall Bottlenose dolphins in superb conditions, going at a nice, easy speed for filming and getting the sound of the dolphins surfacing too. The 'whuff, whuff, whuff' as three came to the surface in quick succession made the sound guy wave both thumbs in the air – he was loving this as much as anyone.

The dolphins kept us company for well over half-an-hour, by which time we were a good bit further along the coast and had seen the majestic but rather sad ruins of Findlater Castle before we reluctantly turned the boat at Sandend Bay and stopped for a breather, letting the dolphins continue on their way. The producer was in mild shock and tears just kept running down her face as she exclaimed "That was just incredible, the dolphins were SO beautiful, and the BABIES… how on earth did you arrange that Charlie? Wow, just wow… I can't believe that just happened, oh my goodness…" The rest of the film crew were celebrating getting such wonderful footage safely in the can too. It makes such a difference (and can be frustratingly rare) when you are filming unpredictable wildlife and all the bits of the puzzle fit beautifully

Page 80: ID#990 *Flake* in the middle of a scrum with some of the *Bad Boys' Club*.
Page 81 top: a sub-adult dolphin completely inverted above another, with Moray Firth landscape in the background. Bottom: one young dolphin breaching beside the boat with three friends behind her.

together for a change and you end up with exactly the sort of material that you had in your mind's eye when the storyboard was being drawn and the script discussed. We had clear, sharp, beautifully lit footage and the wild and free stars of the show just lining up near the boat to be filmed – perfect. We encountered some more dolphins on the trip back to Buckie that day, a real bonus. As usual with filming a programme, we had to balance up the dolphin footage with some scenic shots of the coastline and then had to fill in one or two bits and pieces of Hermione's script that got a bit lost in the earlier dolphin excitement. Eventually it was a tired, pretty sunburned but extremely happy film crew that disembarked at Buckie Harbour later that afternoon after what had been, without doubt, one of the best filming boat trips in the Moray Firth that I have ever, ever had.

Opposite top left: three dolphins going at full speed chasing after fish. Top right: robust play with three sub-adult dolphins.
Bottom: ID#1113 *Lunar*, son of ID#580 *Moonlight*, to the right of ID#1109 *Puddles*.
Above: ID#969 *Spirit* with her baby surfacing alongside her.

Chapter 20

Voices in your head

It is funny how little voices in your head can sometimes persuade you to do the exact opposite of what you intended. After dropping my wife off at our Art and Craft Studio at North Kessock, I decided to stay in the village for the rising tide, which was imminent. It is high summer and the sightings of dolphins have been good everywhere so the salmon run must be at least very reasonable if not brilliant. I had intended going to Chanonry Point for the day but this particular morning I made an executive decision to have a more peaceful day on the old sandstone ferry slipway opposite Carol and Dave Greer's house. Apart from enjoying the tranquillity and being able to concentrate on what I am doing with the dolphins, the only extra task is to phone up to my colleagues at the Dolphin and Seal Centre approximately 30 metres directly above me on the A9 car park and inform them of which individual dolphins are out in the water hunting for food. Hopefully, one or two of our adoption dolphins will appear as a bonus. Being able to talk to them up in their elevated vantage point above the Kessock Channel should mean that if I lose sight of the dolphins I can quickly find out in which direction they've gone... very handy!

Setting up my camera gear is done pretty much by instinct these days; I have been doing this for so long now and have a fairly simple method and system in place that before I know it everything is switched on ready to go. Not long after taking my first couple of frames to check exposures, I hear a dolphin venting not far from the slipway and sure enough Kesslet comes gliding towards me with Charlie, Zephyr and young Breeze in close formation. There hadn't been many big groups in the Kessock Channel for a few weeks so this is a reasonable start. The tide doesn't seem to be moving very quickly yet but it is one of these very calm days when you can't really judge the tide's speed. Indeed, it is deceptive because Kesslet strikes a fish immediately she enters the current

in the middle of the Channel and is carried back with it for a few metres until she regains control of her position, struggling with the huge salmon that, unusually, is still putting up quite a fight. A different sub-group of dolphins comes racing in from the direction of the Kessock Bridge – a collection of mums and calves including Moonlight, Rainbow and Sickle with their youngsters following closely behind. Sundance, Scoopy, Mischief and Denoozydenzy come in from the other side of the tide representing the fully grown 'lads' and they take up various strategic positions in the, by now, very fast incoming current.

The water underneath the dolphins in the middle of the Kessock Channel is over 30 metres deep and the seabed has a very lumpy profile, causing all sorts of whirlpools and eddies which make it fascinating to sit and watch for a while. The dolphins have learned to judge where the incoming fish are running and sit swimming against the current, pretty much as they do at Chanonry, expending the minimum of energy for the maximum return of protein and calories. I had already phoned up to the Dolphin Centre to say who I could identify so far but they phoned me back to say that they could see more dolphins heading my way from the harbour area, closing in fast. Sure enough, some of the sub-adults are hammering towards the other dolphins, half breaching out of the water as they approach, overshoot them and about-turn quickly (the dolphin equivalent of a 'handbrake' turn) to set up their own little hunting party fifty metres or so behind the main group so as not to antagonize any of the adults.

For the next three hours the dolphins hunted really well, with some fine fish being thrown in the air and lots of chasing about. More dolphins arrived after a while including Midnick, Spirit and Sparkle plus a few others. For one glorious day I had all six adoption

dolphins in the Kessock Channel parading around in front of me at the one time! A few people who had been up at the Dolphin Centre came and stood with me, all WDC supporters, some of whom had adopted dolphins and I was able to show these nice people through the big lens their very own dolphin, wild and free the way they all should be. A magical, stress-free day. What put a funny twist on it was that according to some of the regular watchers, very few dolphins had stopped at Chanonry Point to hunt – they all passed through heading for Inverness and the Kessock Channel and never really did much at Chanonry at all. It's strange how things can work out sometimes.

It can be very interesting, indeed fascinating from a people-watching point of view, seeing the reaction when people have just seen our dolphins for the first time. Some people just keep smiling and saying "Wow…oh wow!" while others jump up and down squeaking and babbling incoherently. Some people are simply stunned by the fact that you can see these totally wild animals only a few metres in front of you while you are standing on a beach, while others are bowled over by their sheer size and physical bulk.

I remember a nice wee lady grabbing my jacket sleeve excitedly one day as big Thunder went past – he was an enormous old guy with a towering dorsal fin and the lady said to me "I didn't think you got Killer whales here", convinced that Thunder was actually an Orca. It took a bit of persuading from me to get her to think otherwise. Like the producer out on the *Gemini Explorer* to whom I referred earlier, some people burst into tears after a close encounter and they can't actually explain why. They are tears of happiness of course, but isn't it strange how we can suddenly just 'let go' at the sight of a dorsal fin or two? Don't get me wrong, pretty much every time I see one of 'my' dolphins as I tend to call them, my pulse quickens and my mood changes to 'Very happy – thank you very much' almost instantly, as I'm not immune to the 'Dolphin effect' either, even after watching them for so long. I remember many years ago meeting a man on Cullen beach who apparently just had a blazing argument with a relative whom I knew and was feeling very low about it. But after watching some dolphins surfing in Cullen Bay with me, he went back and sorted things out and, by all accounts, was a much cheerier individual from that day on.

Chapter 21

Education

Every year I really enjoy hosting a group of international students and their lecturer, Dr Steve Kett from Middlesex University, who come up here to do some field work as part of their biology course, which includes getting a talk from me about the dolphins while we stand on the beach at Chanonry. The reaction that I get and see from these young people as the first dolphins begin to appear right in front of them is amazing, always full of wonder, joy and genuine appreciation. After that initial sighting the questions just come out of them like an avalanche and I know from the subsequent feedback that I get from Steve that the dolphins are very much the highlight of the trip. I get e-mails, letters and texts from many people every year, either WDC supporters and/or holidaymakers after having been up here dolphin-watching, some for the first time. They sometimes go into great length about their visit or holiday, telling me every little detail and maybe send me photos to see if I can identify which dolphin or dolphins they saw. I love getting this feedback and helping out even though it can take up quite a bit of time. I find it very satisfying to know that in our always switched on high-tech society, never-ending social media intrusion and inane reality television programmes, that the seemingly simple, low-tech act of standing on a beach or sitting in a tour boat to see these wild, free dolphins can still have such a deeply profound effect on us. I sometimes arrange to meet school parties at Chanonry to help children (of all ages) learn about the local wildlife outside of the classroom. Kids are great at seeing right through things with no pre-conceived ideas and some of the questions that they ask are incredible. They certainly keep me on my toes while I'm with them, and you can't fob them off with vague answers either. A rather serious looking young girl from one of the local schools said to me one day, "That's Rainbow the dolphin over there – do you think she has caught a fish?" I asked her how she could tell it was Rainbow, from such a distance, not telling her that it was me that did the text and photos for 'Adopt a Dolphin' updates. She snorted at me, "By her dorsal fin of course, don't you know anything?" That was me put well and truly in my place!

Chapter 22

Food, glorious food

Bottlenose dolphins are what we call opportunistic feeders and although, like us, they have favourite types of food such as salmon and mackerel, they also eat things that we don't often get to see them catch as this hunting usually happens out of our sight. We know that they have quite broad diets through analysis of stomach contents taken from dead animals that have been recovered for post mortem examination. The veterinary pathology staff at Scottish Rural College (formerly Scottish Agricultural College) in Inverness undertake this and thanks to them and the collection of bones, fish ear-stones (or otoliths as they are known) and other stomach contents we know fairly well from what 'menu' our dolphins have to choose. As the Inner Moray Firth narrows increasingly from Chanonry Point to the Kessock Bridge and west into the Beauly Firth, the edges of this big estuary become quite shallow. There are fish types that can be an alternative when the protein-rich salmon and sea trout are in short supply, such as flatfish and eels. I have seen and photographed our lovely resident otters catching and eating these fish here too, but for a big wholly aquatic mammal like a dolphin, trying to catch this type of food has physical as well as habitat challenges. As the water in which to hunt becomes shallower, one mistake by an over-ambitious dolphin can lead to it becoming stranded, especially on an ebb tide.

I have witnessed dolphins hunting flounders in the flat and clear sandy bottom of Cullen Bay by looking over the side of the boat through the surface of the water with a small seaside plastic bucket that had the base cut out. I could see the dolphins whizzing along only a few centimetres from the sea bed and then suddenly snatching the flatfish out of the sand. It has been demonstrated that dolphins in the tropics do this by 'buzzing' the sand with intense bursts of echolocation clicks and this action scares the fish out of the sandy silt, so there was no reason to believe that our dolphins were doing anything different.

This isn't the only time that I have witnessed our dolphins doing something that has been described as 'new' or 'unique' on wildlife programmes. One day as I was driving down the shore road of the Beauly Firth from Redcastle I saw a big splash and then some water spray coming from near the shore. The tide was just under halfway in and, as I couldn't see clearly with the naked eye what was happening, I stopped the car and quickly grabbed my binoculars. Instead of being an otter or a seal chasing something through the shallows, which at first I thought it was, it turned out to be a dolphin racing through the water at top speed, water that was way less than a metre deep. Coming in this close to the shore seemed almost suicidal for the dolphin and a quick glimpse of the very recurved dorsal fin told me that it was Kesslet 'hydroplaning' through the shallows and grabbing fish as she sped along. And I mean sped – she was going at full throttle, so fast that her young son Charlie could hardly keep up. I could see now that she was hitting quite reasonably sized disc-shaped fish, possibly small skate or flounders that lay in her path, skilfully snatching them in her mouth before skimming round into deeper water to eat them. This was an ingenious but heart-stoppingly scary method of hunting to my way of thinking. All I could do was watch as young Charlie did exactly the same as his mother with the same result, except he wasn't getting anything as common as flounders, oh no – he struck an absolutely beautiful sea trout and grabbed it by the head before going out to join Kesslet who was polishing off her catch, no doubt showing off to her that anything she can do, he can do better. Kids, eh?

Opposite: ID#23 *Mischief* throwing a large salmon in the air with the Easter Ross coastline in the background.

ID#440 *Sickle* with a lovely fish for lunch at Chanonry Point, Black Isle.

I have not seen dolphins catching eels that often, but I did see old Sutor having a long tussle down near the Kessock Bridge one day with a huge eel, possibly a conger, but it kept wrapping itself round and round his jaws making it very difficult to swallow. I could hardly take photos for laughing at this comical scene – not that funny for the eel mind you, but it was a full twenty minutes or so before Sutor managed to get it untangled from his jaws and successfully swallowed it with what could have been a look of utter relief on his face. Having only pectoral fins that are not much help in holding or grasping prey can be a bit of a disadvantage in such situations. Otters on the other hand have strong and dexterous front paws with sharp claws, ideal for coping with such long, wriggly and

difficult prey. But dolphins have to rely on very powerful jaws fitted with around one hundred teeth in total, teeth that are called 'homodont' – pretty much all one shape and size which are really only used for grasping and crushing the heads of prey such as fish and squid, then the fish are swallowed using a combination of the very powerful tongue and throat muscles. Omnivorous mammals, like us humans for example, have 'heterodont' teeth which are mixed shapes and sizes such as canines, incisors and molars which are better suited to cutting through meat and tearing food apart. Dolphins therefore have to be careful about what they catch as it could be a total waste of energy to acquire something that they cannot physically process.

Food sources for dolphins and other marine mammals are not always concentrated into one convenient small area, like the Chanonry and Kessock Narrows for instance. Travelling to find food is an energy-sapping but necessary activity for dolphins in this population. Bottlenose dolphins, especially these physically large and heavy eco-types in the chilly waters of the North Sea have to maintain their thick, insulating layer of blubber (around twenty centimetres in places) and can theoretically go without food for a day or so if they really have to. However, this is a precarious and dangerous situation – eating into their precious lipid reserves by not catching enough to eat daily not only lessens the effectiveness of their thermal 'jacket', making them more and more vulnerable to the cold but also introduces the likelihood of metabolising any ingested toxins. This can be really nasty, insidious stuff that has been taken on board through contaminated prey, substances like heavy metals and polychlorinated biphenyls that have been stored fairly inertly in the animals' blubber layer, possibly for many years, but once these toxins mobilise themselves around the dolphins' bloodstream and enter vital organs and tissues – it's bad news. Something that is sometimes forgotten is the sad and cruel fact

that mothers can pass a lot of this stored contamination directly down to their babies through the milk supply. This is possibly why the first calf survival rates of dolphins that forage on contaminated prey species can be so low.

Dolphins can travel relentlessly day and night if needs be to find food and, as fish stocks fluctuate (sometimes to near extinction levels) some of our dolphins have decided to search out pastures (and larders) new. They will cover a long distance in a relatively short space of time, hence the large percentage of recognisable dolphins now seen down the east coast of Scotland and indeed, over the border into England on a pretty regular basis. Grabbing food on your travels is an important part of being a free-ranging top marine predator and not being too fussy about your diet can sometimes mean the difference between life and death, especially when times are seasonally lean. Cod, whiting, mackerel, saithe, haddock, salmon, sea trout, herring, sprats, plaice, squid and octopus are amongst the food remains found in some of our dolphins' stomachs, which shows the variety of prey species that can be available from season to season. We don't expect to see much other than salmon and sea trout being caught around the Cromarty Firth entrance, Chanonry Point and the Kessock Channel as most people watch dolphins in the spring/summer season. But if we could somehow follow dolphins around over the course of a year without disturbing them, then we could see the sheer variety of food that they can eat. We would also witness some of the fascinating and very clever hunting methods, singly and as a group that they use to catch their prey. Imagine if you and your friends should decide to 'up sticks' and go in search of quieter territory that has a good food supply and doesn't belong to anyone (or anything) else and after a week or so of exploring you come to like an area that you have discovered and think to yourself, this will do nicely. Those are possibly the reasons for seeing the expansion of the known range of our dolphins these days: fundamental requirements for life, food and somewhere to live.

A typical day's dolphin watching at the likes of Chanonry Point during the spring and summer will normally involve seeing dolphins using the 'against current foraging' technique of hunting salmon and sea trout that I described earlier in the introduction to this book. No matter how often I see dolphins using this method,

Opposite top: ID#866 *Zephyr* manipulating a fish to make it easier to swallow, Chanonry Point. Bottom left: a dolphin forcefully ejecting a fish to re-swallow more safely. Bottom right: a salmon face to face with its Nemesis, ID#580 *Moonlight*. Above: ID#866 *Zephyr* re-swallowing a fish after ejecting. Dolphins may do this many times until it is comfortable to swallow.

I always marvel at the apparent food-on-a conveyor-belt simplicity of it. Sometimes, if the fish are smaller and are running deeper in the tidal current, then we might not see the actual moment of capture or ingestion by the dolphin. On the other hand, if the salmon or sea trout are a bit bigger, this is where some of the more spectacular visual dramas occur, with fish being struck clean out of the water, sometimes many metres into the air. If you manage to capture this on camera you can get an idea of just how big these dolphins like their prey to be – mind you, Bottlenoses are nothing if not enthusiastic and some individuals, like Kesslet, have a habit of attacking fish that are ridiculously large in comparison to their own body size and can easily take over half-an-hour to consume comfortably.

Sometimes, and this varies from dolphin to dolphin, the larger fish that have been caught may not fit comfortably in the throat as swallowing commences and the fish has to be regurgitated, manipulated and re-positioned until the dolphin is confident that nothing is hampering the progress down towards the first stomach. This could not be successfully achieved if the fish was still alive and possibly make its escape, so the dolphin has the fish immobilised immediately by biting hard at the base of the skull or crushing the head itself. Some dolphins prefer the fish to be, as

we would see it the 'right way up', whereas others will position their fish 'upside down' to swallow, but all dolphins have to ingest their fishy dinners head first otherwise gills and fins could jam in the throat and become stuck. Unlike you or me, dolphins can't simply go to the fridge and take out a sandwich if they get peckish. They have to expend valuable energy to catch something and when the opportunity for hunting arises they have to consume as many fish as possible, as they can never take it for granted that their next meal will be either soon or easy to catch and kill. There is a balance to be struck however; the energy budget of any wild animal generally governs just how much effort is worth expending in return for the amount of calorific goodness gained from the food itself. There doesn't seem to be much point in pursuing a tiny sprat for over half-a-kilometre, but to exert that amount of energy for the reward of a five-kilo salmon makes much more economic sense. The reason we think that these dolphins have invented and perfected the 'against current foraging' technique is that the expenditure of energy to catch the food is very much less than the hot pursuit technique as the prey are literally swimming towards the dolphin's open mouth, normally with minimal chasing involved unless the fish makes a rare escape. All they have to do is move their tail up and down as much as is needed to stay in the one place in the tidal current. Clever… very clever.

Chapter 23

Babies

One of the loveliest things to see in the natural world is a mother dolphin swimming along in tight formation with her baby. After a gestation period of 12 months or so, the fully developed and ready to swim, metre-long neonate is born tail first and the first important task is to get it to the surface to take his or her first breath of air. Sometimes female dolphins have a 'midwife' with them, perhaps a friend, sister or even her own mother to assist in the birthing process and they will sometimes help the young dolphin to the surface in case instinct doesn't immediately kick in. We only know of single babies being born to females in this population, as in most groups of Bottlenoses around the world that have been studied – but the basic equipment is there to support twins: two mammary slits and the physical capacity to produce enough milk for two. The baby learns quickly to latch onto one exposed nipple that has been pushed through the watertight mammary slit and suckle. A waterproof seal is made around the mother's nipple through frilly extensions at the sides of the baby's tongue. These feathery-looking protuberances on the tongue eventually disappear after a few years, after weaning has taken place.

Opposite: ID#433 *Kesslet* with her first baby in 2007, later to be named *Charlie* after the author.
Above top: ID#732 *Tall Fin* with a year-old baby travelling through the Chanonry Narrows. Bottom: ID#1030 *Crockett* with a year-old baby.

In contrast to the adults or older calves, new babies need to develop a smooth surfacing and breathing pattern and for the first few days of life can often look a bit uncoordinated and flop about when coming up to breathe. This soon evens out and after a while the baby can surface smoothly and in unison with mum. The female has to adjust her normal routine and activity habits to suit what the baby can cope with, but has to balance this with keeping herself sustained with good-quality food to be able to provide the thick, fat-rich milk on which the baby dolphin will thrive and put on weight. It is surprising just how quickly the little dolphin fits in with what mum is doing and by staying close by her side just behind the pectoral fin (they look like they are glued there sometimes), the young dolphin can get a slight towing effect from the slipsteam which helps minimise its energy expenditure.

While little junior is keeping up with mum, he or she is being taught many things – not least of which is his or her 'name', or signature whistle as it is known and what mum's voice, which is also a high-pitched and variable whistle, sounds like. This is vital, possibly life-saving tuition so that the two dolphins can maintain contact by whistling to each other and in the three dimensional water-world in which they live, listen for and locate each other in sometimes dangerous situations. The young dolphin also has to learn how to 'start up' and use its own wonderful and very powerful inbuilt sonar system, called echolocation. Echolocation clicks have to be generated, which is done by passing and recycling air through bladder-like sacs buried deep within the blowhole. He or she then has to practice controlling the speed and intensity of the clicks, focussing them into a beam of useful pulses so that,

Opposite top left: a young calf surfacing beside ID#818 *Yin-Yang* showing yellow lesions on the pale skin. Top right: a pair of adult females with their small calves, each one about a year old. Bottom left: an inquisitive baby checking out the boat the author is on. Bottom right: flying baby – it doesn't take dolphin babies long to learn to breach. Above: ID#732 *Tall Fin* with tiny newborn calf alongside her.

as the amplitude is varied, the detail of the sound 'picture' increases, as does the effective range of prey detection. These clicks are concentrated by an organ situated in the forehead called the 'melon'. These sound pulses are passed out into the water, hit any object within range and bounce back to be received by the oil-filled lower jaw that is hard-wired into the auditory system, which is a very effective reception aerial. Very soon, with a lot of practice, the young dolphin, who is now looking like a scale model of mum, will be able to determine shapes and sizes, textures and densities of things in the surrounding sea. Another crucial capability is to be able to track and calculate just how fast something is approaching – like a fish for example, once the young dolphin has moved from milk to solid food at around eighteen months of age.

The young dolphin accompanies mum everywhere for the first six months or so and rarely leaves her side. Babysitting happens fairly frequently but only with trusted friends or relatives. Being with auntie or granny is great fun and very exciting for a while but it won't be long before the handover happens and back to mum's side we go and back to "Don't do that" and "Stay away from that thing" and so on. Young dolphin calves of any species are vulnerable to

predation and possible attack by sharks or other marine mammals such as Orca, but it is known that some male Bottlenose dolphins will attack and sometimes kill young dolphins to try and make the female become sexually receptive. Female dolphins have a strategy of mating with multiple males to make attacks rarer and she can possibly infer to the male dolphin that wants to attack the calf that he could well be the father – a clever strategy, but sadly it doesn't always work.

As the young dolphin grows, which appears to happen in growth spurts, it will not only be learning valuable life skills but is also finding its place and position in a socially complex society that has rules and a pecking order, and discipline is often used to bring an errant calf back into line. As mum begins to reduce the milk supply, normally by about eighteen months or so, hunting tuition increases and the calf is increasingly encouraged to start catching small fish to supplement the mother's dwindling reserves of milk. Just like human children, some young dolphins learn quickly while some others might take a while to get the hang of capturing and dealing with their own food. Many a calf I have watched over the years, catching their first tasty little fish and then carrying it around showing it off to mum, friends and extended family before swallowing it with great satisfaction. You are now a top marine predator, young one!

Opposite: ID#1027 *Hawkins* with youngster surfacing beside her.
Page 104 top: ID#31 *Rainbow* and neonate baby at Chanonry Point. Bottom: ID#433 *Kesslet* with her second baby, only hours old, under the Kessock Bridge.

Page 105: another inquisitive baby checking out the boat the author is on.

Chapter 24

Messing about in boats

Nobody likes being out on the water on-board a boat more than I do – let me get that confession out of the way right now. What we often forget when we humans put a boat in the water is that we are entering the dolphins' home, and uninvited at that. If we tiptoe quietly and politely around and don't go racing about making a horrendous noise and imposing ourselves on groups of dolphins, hopefully the dolphins will be bothered by us as little as possible. It is a big area of sea out there – the Moray Firth in its entirety is around five thousand square kilometres but the dolphins tend to use only a small fraction of that huge body of water, preferring the narrower channels where the water is deep and flows quickly. This also means that people who are determined to seek out dolphins for themselves with their own leisure craft, instead of going out with one of the well-run and knowledgeable accredited tour operators, have areas that they can target such as the Chanonry Narrows and the Kessock Channel. Over the years I have seen some idiotic and sometimes downright criminal behaviour by people who supposedly love dolphins and have to demonstrate this 'love' by getting as near to them as possible, if at all possible within touching distance. This disrespectful and illegal behaviour isn't just carried out by powered craft like speedboats and jet-bikes either. There is also a particular group of very determined canoe and kayak owners who in recent years seem to want to paddle with dolphins and will not accept any advice that this is intentional disturbance of a protected animal. It disrupts the dolphins' feeding efficiency, can separate mothers and babies and is against the law. I had a very friendly meeting with staff and instructors from the Glenmore Lodge outdoor centre about their organised paddle groups avoiding areas used by dolphins and it was well received and we have had no incidents with tutored and guided groups since. The only problem is with these steely-eyed, determined and seemingly totally deaf individuals (they get shouted at a lot by the dolphin watchers) or clusters of paddlers that want to sit right on top of dolphins, almost knock their brains out with their paddles and then breathlessly tell their office colleagues on Monday morning what a wonderful, natural experience it was.

Chapter 25

What is it about dolphins?

There are plenty of big and intelligent apex (a term to describe the animal at the top of the food chain) predators in the natural world that are gorgeous to look at but few seem to have the 'draw' that cetaceans, and especially dolphins, have for us humans. Bottlenoses (*Tursiops truncatus*) in particular who, overall, are the most studied but probably the least understood of the delphinids,

have a magnetism that attract all sorts of people from all walks of life. I don't think that anyone can absolutely put their finger on what 'it' – that magical something – actually is. I certainly can't. Could it be the perceived very high intelligence that these mammals have or is it their complex social structure and the way that they look after each other? Is it their shape or size or the fact that they, being mammals, have many body parts in common with us – hearts, lungs, brains, kidneys, wombs and so on? Or is it in the more esoteric realm such as the seemingly endless tales of dolphins helping people in distress, curing people with depression or other illnesses? Or is it maybe even something as simple as – and I'm hesitant to use the 'F' word here – the childhood memories of *Flipper*, the dolphin on the telly all those years ago and imagining what it would be like having this lovely friend that lives in the sea who understands every word that you say?

As I say, I don't know for sure – there is absolutely something about them, no doubt about it, but instead of wandering off into the realms of fantasy when we talk about dolphins, and especially our big, special Bottlenoses up here in the chilly waters of the Moray Firth and the north-east coast – maybe we should be a bit more pragmatic and love and admire them for what they actually are. Wild, free, highly intelligent and charismatic apex predators that have a pretty tough time out there just trying to make ends meet from day to day and who have to hunt – and yes, kill, to survive. Let's clean up our own act and stop pumping all sorts of contaminants into their watery home. Stop the overfishing that is literally taking food out of their mouths and if that isn't bad enough, wrapping them up and drowning them in discarded fishing nets too. Let's quieten down the awful racket that we are making in their realm. Imagine trying to eat your dinner while someone is revving

up a motorbike in the same room or trying to converse with each other over the constant clanging of a huge hammer on a bit of steel.

There is going to be a lot of large-scale development around these waters very soon with large wind farms being built out in the Moray Firth. With that comes extra boat traffic, ports and harbours to be made ready and improved to act as pick up/drop off points for the industry. If this is carefully and sympathetically planned and executed then disturbance to local marine mammals should be minimised – but there will be some effect on our dolphins, no doubt about it. There will also be some effect from the building of many new marinas up and down the coast for leisure craft, that seem to fill up as soon as they are opened. More boats equal more possibility of interactions with dolphins and possible disturbance. Let us try to help them, not just here in Scotland but all round the world, by *not* selfishly throwing ourselves in the water beside them and trying to swim with them (where they have no choice in the matter) or capture and keep them in tiny, filthy pools for our 'Infotainment' or supposed 'vital research/conservation breeding programmes' or worse still, kill them as a tradition, supposedly for food. Tell your MP that this kind of thing is not acceptable in this day and age or better still, become a member of Whale and Dolphin Conservation and we can all shout together as one to get something positive done before it is too late.

Opposite: a young calf surfing through the waves, spray going in a nice 'V' formation.

Chapter 26

Some Light Reading

Books are wonderful things. It seems a rather quaint thing to say in this age of instant and 'always on' internet information, but back then in the late 80s/early 90s I would scour libraries for any books, scientific or otherwise, to do with Bottlenose dolphins and read them cover to cover. There were a few scientific papers and reports from different parts of the world that were accessible and they made interesting reading too. To my mind though, the best was a book of a collection of brilliant papers simply titled *The Bottlenose Dolphin* edited by world-renowned cetacean researchers Stephen Leatherwood and Randall R. Reeves (1990 Academic Press, ISBN: 0-12-440280-1). I devoured every word, photograph, chart and illustration. If you really want to learn about what makes these top predators tick at a fine-scale level – beg, steal or borrow a copy. It can be heavy reading in places if you are not used to scientific paper writing but the contributing scientists, some of whom I have either met or corresponded with over the years, are the very best in their respective fields. Nowadays of course, the internet is a fantastic learning resource and if you can weed out the fact from fiction, then there are tons of information about our own particular and very special population of dolphins.

Aberdeen University's Lighthouse Field Station in Cromarty and the Sea Mammal Research Unit at St Andrews University were the pioneers of Bottlenose dolphin field research in the Moray Firth and the North East of Scotland. Many scientific papers and reports are available from their respective websites to study, disseminate and from which to glean up-to-date facts, data and reports. One of the researchers that I used to meet regularly in the Dolphin and Seal Centre, Doctor (now Professor) Ben Wilson from Aberdeen University's Lighthouse Field Station has a great book simply called *Dolphins* still available (ISBN: 978-1841071633). His collaborative book with Professor Paul Thompson, also from Aberdeen University, entitled *Bottlenose Dolphins* (ISBN: 978-0896585263) tells in one chapter the story of Kesslet's mum – Kess, a very special matriarch with a significant physical disability of the spine. I used to watch her nearly every day in the Kessock Channel and erected a memorial plaque to her at the slipway at North Kessock after she died. Both books by these world class scientists are well worth a read and if you are into dolphins at all – a real 'must have' for your bookshelf.

Acknowledgements

There are quite a few people that I have to thank for helping me (knowingly or not) put this book together.

My parents, Jimmy and Bertha Phillips, both sadly no longer with us, who altered their retirement plans completely to let me go off on this mad dolphin trip that I am thankfully still on.

My wife Susan, of course, for simply putting up with me and supporting me over the years, and enduring the sometimes *very* early morning departures and occasional stupidly late finishes to get the job done and tolerating the car sometimes being a bit whiffy as there may just have been a dead porpoise or seal in it the day before (www.aurorabearealis.co.uk).

My buddy Alan Ward, whose West Country wisdom, wit and friendship and countless texts, e-mails and phone calls about sightings of dolphins if I'm not at Chanonry Point have over the years been so very much appreciated. All the boat trips out have been fun too – here's to many more, my friend.

The staff, past and present, at Aberdeen University Lighthouse Field Station, Cromarty (www.abdn.ac.uk/lighthouse), who have been so hospitable and helpful to me over the years including Paul Thompson, Ben Wilson, David and Susan Lusseau, Tim Barton, Gordon Hastie, Vincent Janik, Ross Culloch, Simon Ingram, Isla Graham and Andy Foote. Last but not least Barbara Cheney, Photo ID Officer, for being such a star, looking at and grading countless thousands of my crappy dorsal fin photos over the years and for trying to get me out on the research vessel *Rona* as often as she can for me to take even more pictures.

Grateful thanks to Simon King OBE for kindly agreeing to write the foreword for this book. You are a busy lad Simon so I really appreciate the time taken out of your hectic schedule. We need to stand and watch dolphins up here together again one day, my friend. (www.simonkingwildlife.com)

To Bob Reid, Andrew Brownlow, Nick Davison and the pathology staff past and present at SRUC (formerly SAC) (www.strandings.org) Inverness for letting me help recover then photograph, record and learn from the post mortem examinations over the years of everything from a diminutive Harbour Porpoise to a mighty Sperm Whale. A procedure gruesome to some, (not me) but utterly fascinating none the less.

My colleagues and friends, past and present at Whale & Dolphin Conservation HQ in Chippenham (uk.whales.org), Scottish Dolphin Centre at Spey Bay and Dolphin and Seal Centre at North Kessock, including my old work buddies from Ross & Cromarty District Council and later Highland Council – too many to name really, but you know who you are.

The very wonderful Sarah Pern and her staff, past and present, at Ecoventures, Cromarty (www.ecoventures.co.uk) for the hundreds of hours spent out on the water (for free) over the years on board the R.I.B. *Saorsa* and the many great encounters with these amazing dolphins that we have shared. Here's to many more.

To the other commercial boat trip operators in the Moray Firth that have been so kind over the years, whether helping out with filming, giving advice or having me on board to try and get some photos of mums and new babies or even stray whales – thanks to Pippa and Simon at North 58 Adventures at Findhorn (www.north58.co.uk), Eric Wardlaw of Phoenix Cruises, Inverness (www.phoenix-boat-trips.co.uk), Gwyn and Paula Tanner of Dolphin Trips Avoch (www.dolphintripsavoch.co.uk), Bill Ruck of Moray Marine at Lossiemouth (www.moraymarine.com/welcome/about_us.asp) and Iris and Davey Smith of *Gemini Explorer* (www.geminiexplorer.co.uk) in Buckie – long may you all keep afloat!

Canon (www.canon.co.uk), for making such high quality, robust equipment with which I have earned my livelihood and the friendly and knowledgeable staff of Ffordes of Beauly (www.ffordes.com) for taking so much money from me over the years.

Minox UK (www.minox.com/index.php?id=9316&L=1) for supplying me with their wonderful binoculars and scopes – makes the job of spotting dolphins at a distance so much easier.

Last, but not least I would like to thank Colin and Eithne Nutt of Ness Publishing (www.nesspublishing.co.uk) for having the vision, clarity of thought and confidence in me to take this book forward and to actually make it happen, and to Louise for her careful thought and attention to detail with the book's layout and design.

And, most importantly, to the stars of this book… to all of our wonderful, wild, free gorgeous dolphins, past and present – I love you all, each and every one.

All dolphin photographs in this book are of wild, free-ranging Bottlenose dolphins taken in the Moray Firth and are all strictly © Charlie Phillips/WDC or © WDC/Charlie Phillips or © Charlie Phillips Images. All rights reserved.

Published 2015 by Ness Publishing, 47 Academy Street, Elgin, Moray, IV30 1LR
Phone 01343 549663 www.nesspublishing.co.uk

Design by Louise at chatterboxdesign@mail.com

ISBN 978-1-906549-90-9

Front cover:

ID#1025 *Charlie*, son of ID#433 *Kesslet*, one of the Firth's resident Bottlenose dolphins bursting through the water towards the author's camera. The author of this book had the honour of having this young dolphin named after him by Aberdeen University's Lighthouse Field Station.

Rear cover:

ID#1109 *Puddles* looking up at the author's camera from a metre or so underwater – the water in the Moray Firth is not normally as clear as this.

ID#1126 *Doyle*, daughter of ID#732 *Tall Fin*, breaching from blue water in the Inner Moray Firth.